A-LEVEL YEAR 2

STUDENT GUIDE

OCR

Biology A

Module 6

Genetics, evolution and ecosystems

Richard Fosbery

HODDER
EDUCATION
AN HACHETTE UK COMPANY

Hodder Education, an Hachette UK company, Blenheim Court, George Street, Banbury, Oxfordshire OX16 5BH

Orders

Bookpoint Ltd, 130 Park Drive, Milton Park, Abingdon, Oxfordshire OX14 4SE

tel: 01235 827827

fax: 01235 400401

e-mail: education@bookpoint.co.uk

Lines are open 9.00 a.m.–5.00 p.m., Monday to Saturday, with a 24-hour message answering service. You can also order through the Hodder Education website: www.hoddereducation.co.uk

This guide has been written specifically to support students preparing for the OCR A-level Biology A examinations. The content has been neither approved nor endorsed by OCR and remains the sole responsibility of the author.

Cover photo: Argonautis/Fotolia; p.66: Fix the Fells, Lake District National Park Authority

Typeset by Integra Software Services Pvt Ltd, Pondicherry, India

Printed in Italy

Hachette UK's policy is to use papers that are natural, renewable and recyclable products and made from wood grown in sustainable forests. The logging and manufacturing processes are expected to conform to the environmental regulations of the country of origin.

Contents

■ Getting the most from this book

Exam-style questions

Commentary on the questions

Tips on what you need to do to gain full marks, indicated by the icon ⓔ

Sample student answers

Practise the questions, then look at the student answers that follow.

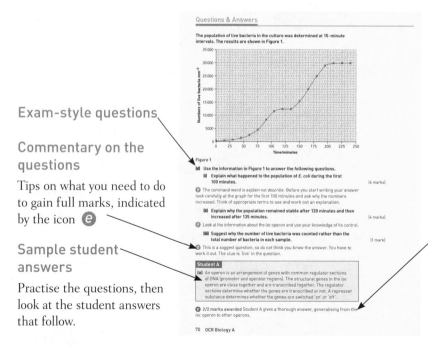

Commentary on sample student answers

Read the comments (preceded by the icon ⓔ) showing how many marks each answer would be awarded in the exam and exactly where marks are gained or lost.

■ About this book

This guide is the fourth in the series covering the OCR A-level Biology A specification. It covers Module 6: Genetics, evolution and ecosystems, and is divided into two sections:

■ The **Content Guidance** provides key facts and key concepts, and links with other parts of the A-level course. The synoptic links should help to show you how information in this module is useful preparation for other modules.

■ The **Questions & Answers** section contains two sets of questions, giving examples of the types of question to be set in the A-level papers. The first of these — Paper 2: Biological diversity — has some multiple-choice questions and some structured questions, together with answers to all the structured questions written by two students. There are comments on all the answers. The second paper — Paper 3: Unified biology — has some questions that set the topics in Module 6 in a wider context.

This guide is not just a revision aid. You will gain a much better understanding of the topics in Modules 2 to 6 if you read around the subject. I have suggested a few websites that you can use for extra information. These will help especially with topics that are best understood by watching animations of processes taking place. As you read this guide remember to add information to your class notes.

The **Content Guidance** will help you to:

■ organise your notes and to check that you have highlighted the important points (key facts) — little 'chunks' of knowledge that you can remember

■ understand how these little 'chunks' fit into the wider picture; this will help to support:

  ■ Module 5, which is covered in the third student guide in this series

  ■ Modules 2, 3 and 4, which you are most likely to have studied in the first year of your course; these are covered in the first two student guides in this series

■ check that you understand the links to the practical work, since you must expect questions on practical work in your examination papers. Module 1 lists the details of the practical skills you need to use in the papers

■ understand and practise some of the maths skills that will be tested in the examination papers. Examples are indicated by this icon:

The **Questions & Answers** section will help you to:

■ understand which examination papers you will take

■ check the way questions are asked in the A-level papers

■ understand what is meant by terms like 'explain' and 'describe'

■ interpret the question material — especially any data that you are given

■ write concisely and answer the questions that the examiners set

Content Guidance

▉ Cellular control

Key concepts you must understand

In Module 2 you learned that genes code for polypeptides. These are the **structural genes** that are expressed in the phenotypes of organisms. There are many other lengths of DNA that have other functions, such as regulating the expression of the structural genes. These are known as **regulatory genes**.

Almost all cells have a complete genome of the species to which they belong. Some of these genes are required by all cells, but in multicellular organisms a very large number are only required by specialised cells. Some of these genes are expressed in a few cell types; sometimes just one. When genes are expressed they are transcribed and translated during protein synthesis, as shown in Figure 1. The details of protein synthesis are in the first guide in this series.

Exam tip

The first four sections of Module 6 require a good knowledge of the structure of DNA. It is especially important that you understand the antiparallel arrangement of the two polynucleotide strands.

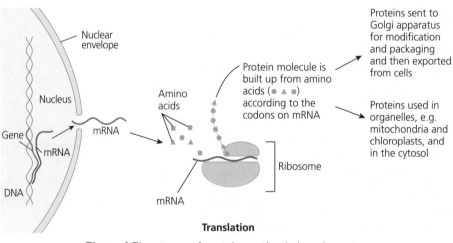

Figure 1 The stages of protein synthesis in eukaryotes

Knowledge check 1

How would a diagram showing protein synthesis in prokaryotes differ from Figure 1?

DNA is a store of genetic information for the synthesis of polypeptides. The sequence of bases in a length of DNA determines the sequence of amino acids in the primary structure of a polypeptide. Each triplet of bases codes for an amino acid. There are four bases in DNA (A, T, C and G) so it is possible to make 64 different triplets. DNA is a template for making messenger RNA (mRNA), which conveys short-lived copies of the base sequence from the nucleus to ribosomes in the cytoplasm.

Angiotensin 2 is a short peptide hormone that has a powerful effect on increasing blood pressure. It is formed from the larger protein angiotensinogen (AGT). Figure 2 shows how the sequences of the bases in the two polynucleotides within the gene for

Exam tip

The stages of protein synthesis are in Module 2. If you do not recall the details, now is a good time to revise that section thoroughly before reading on.

AGT that code for angiotensin 2 relate to the sequence in mRNA and the anticodons in tRNA molecules that identify the amino acids for translation.

5′	GAC	CGG	GTG	TAC	ATA	CAC	CCC	TTC	**3′**
3′	CTG	GCC	CAC	ATG	TAT	GTG	GGG	AAG	**5′**
Codons on mRNA	GAC	CGG	GUG	UAC	AUA	CAC	CCC	UUG	
Anticodons on tRNA	CUG	GCC	CAC	AUG	UAU	GUG	GGG	AAG	
N-terminal	Asp (D)	Arg (R)	Val (V)	Tyr (Y)	Ile (I)	His (H)	Pro (P)	Phe (F)	**C-terminal**

Figure 2 The base sequences for DNA and RNA that are involved in the synthesis of the peptide hormone angiotensin 2

The polynucleotide along which the sequence of RNA nucleotides is assembled is the **template strand** for transcription; the complementary polynucleotide is the **coding strand**. As you can see, the coding strand has a sequence of bases that is the same as that of the mRNA except that in mRNA U replaces T.

The genetic code can be written as the triplets of bases in DNA or the triplets in the coding strand of mRNA. Table 1 shows the 64 mRNA codons. Throughout this book you will need to use this table. If you require DNA triplets then convert U to T — for example, the DNA triplet for methionine on the coding strand is ATG (on the template strand it is TAC).

Table 1 The genetic code; amino acids are identified by using three letters and one letter as shown

		Second position				
		U	**C**	**A**	**G**	
First position (5′ end)	**U**	UUU ⎤ Phe (F) UUC ⎦ UUA ⎤ Leu (L) UUG ⎦	UCU ⎤ UCC ⎥ Ser (S) UCA ⎥ UCG ⎦	UAU ⎤ Tyr (Y) UAC ⎦ UAA stop UAG stop	UGU ⎤ Cys (C) UGC ⎦ UGA stop UGG Trp (W)	U C A G
	C	CUU ⎤ CUC ⎥ Leu (L) CUA ⎥ CUG ⎦	CCU ⎤ CCC ⎥ Pro (P) CCA ⎥ CCG ⎦	CAU ⎤ His (H) CAC ⎦ CAA ⎤ Gln (Q) CAG ⎦	CGU ⎤ CGC ⎥ Arg (R) CGA ⎥ CGG ⎦	U C A G
	A	AUU ⎤ AUC ⎥ Ile (I) AUA ⎦ AUG Met (M)	ACU ⎤ ACC ⎥ Thr (T) ACA ⎥ ACG ⎦	AAU ⎤ Asn (N) AAC ⎦ AAA ⎤ Lys (K) AAG ⎦	AGU ⎤ Ser (S) AGC ⎦ AGA ⎤ Arg (R) AGG ⎦	U C A G
	G	GUU ⎤ GUC ⎥ Val (V) GUA ⎥ GUG ⎦	GCU ⎤ GCC ⎥ Ala (A) GCA ⎥ GCG ⎦	GAU ⎤ Asp (D) GAC ⎦ GAA ⎤ Glu (E) GAG ⎦	GGU ⎤ GGC ⎥ Gly (G) GGA ⎥ GGG ⎦	U C A G

Third position (3′ end)

Genetic code Sequences of three bases (from A, T, C and G) in DNA and mRNA that code for the 20 amino acids in proteins.

Codon A triplet of bases that codes for an amino acid. There are three STOP codons that do not code for any amino acid, but signify the end of a coding sequence.

Exam tip

You will need to refer to the genetic code in Table 1 to answer knowledge check and exam-style questions on cellular control and manipulating genomes.

Exam tip

It is a good idea to learn a few codons to use as examples in your answers. You need to know how to use the genetic code, but you do not have to memorise all of it!

At this point it is a good idea to revise the structure of proteins. Figure 3 shows the main points you should recall from Module 2. In addition, you should know that the term **domain** is used to describe part of a globular protein that serves a particular function.

Knowledge check 2

The base sequence TATGGCGGTTTCATG on the coding strand of DNA codes for the hormone met-enkephalin. Use Table 1 to explain fully why the sequence TATGGGGGCTTCATG also codes for met-enkephalin.
....

Exam tip

The genetic code consists of the codons shown in Table 1 (e.g. AUG = methionine/M) and the complementary DNA triplets. It is *not* the sequences of bases that code for sequences of amino acids as shown for met-enkephalin and angiotensin 2.
....

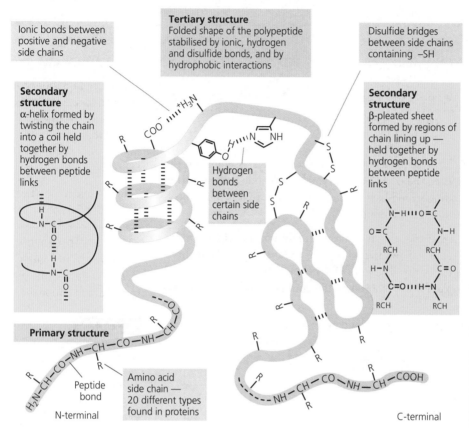

Figure 3 The primary, secondary and tertiary structure of a globular protein

Gene expression

Genes encode proteins and proteins dictate cell function. Therefore, the thousands of genes expressed in a particular cell determine what that cell can do. Each stage in protein synthesis provides an opportunity for cells to regulate their activity by controlling the types of protein produced and the quantity of each type.

Some genes are transcribed and translated all the time; they do not require any external stimulus. Examples are the genes that code for enzymes that are active all the time, such as those that catalyse the reactions of respiration. Other genes are only expressed when there is some form of stimulus, such as a hormone or the presence of an enzyme substrate, to stimulate transcription. It would be a waste of energy and materials to make the proteins encoded by these **inducible genes** if the cells do not need them.

In prokaryotes, such as bacteria, genes that control the same feature are often regulated in groups. Each of these regulation units is an **operon**. The control of gene

Operon A group of structural genes that share the same operator and promoter regions. The structural genes are transcribed together.

expression in eukaryotic cells is more complex. You need to know about these three levels of control:

- transcriptional control
- post-transcriptional control
- post-translational control

Genes change as a result of damage and misreading during replication. These changes are **mutations** and they can occur at the level of the chromosome and at the level of the gene. The smallest mutation is a change to one base pair in DNA, which is an example of a gene mutation.

Key facts you must know

The *lac* operon

Lactose is a disaccharide sugar composed of the monosaccharides glucose and galactose held together by a glycosidic bond. The bacterium *Escherichia coli* lives in the human gut. It is capable of breaking down lactose by breaking the glycosidic bond and respiring the glucose and galactose. *E. coli* lives in the large intestine in mammals and most lactose is digested in the small intestine. However, some may pass undigested to the large intestine. *E. coli* is also found on the udders of cows, so can absorb lactose from milk that remains there.

The enzyme that catalyses the hydrolysis of lactose is α-galactosidase. It is an **inducible enzyme**. The synthesis of the enzyme is controlled along with two other proteins that are required for the absorption and metabolism of lactose. These three structural genes are transcribed together, giving one mRNA transcript.

> **Exam tip**
>
> It is a good idea to follow an animation as you read about the *lac* operon. You will find plenty of animations online that show this method of gene expression. One of them is at www.dnaftb.org/33/animation.html.

Figure 4(a) shows how these genes are 'turned off' when there is no lactose in the surroundings. However, when lactose is present it enters *E. coli* and, if there is no glucose available at all, activates the genes so that they are transcribed and the three proteins are produced (Figure 4(b)). The operon consists of the following:

- a **promoter region**, where RNA polymerase binds to start transcription
- an **operator region**, where the repressor protein binds
- **structural genes** for α-galactosidase, a membrane transport protein for lactose and another enzyme, transacetylase

In addition to the *lac* operon, there is a **regulator gene** elsewhere on the bacterial chromosome that codes for a repressor substance that binds to the operator region, inhibiting transcription of the three *lac* genes *z*, *y* and *a* when lactose is not present. The repressor is an allosteric protein with two binding sites: one for the DNA of the operator region and the other for lactose. When lactose binds the protein changes shape so can no longer bind to DNA. Lactose binds to the repressor and this 'turns on' the *lac* operon.

Mutation A change to a chromosome (chromosome mutation) or to a gene (gene mutation).

> **Exam tip**
>
> Chromosome mutations involve increasing or decreasing the number of chromosomes in the nucleus, or a change in chromosome structure. They are not required knowledge for the examination.

Promoter region A sequence of DNA that controls the expression of one or more structural genes as it is the site of binding of RNA polymerase.

Regulator gene A gene that codes for a regulatory protein that binds to the operator region.

(a)   No lactose present

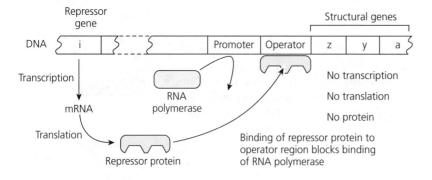

(b)   Lactose present

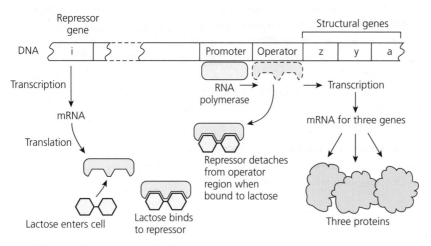

Figure 4 How the *lac* operon functions (a) when lactose is absent and (b) when lactose is present and there is little or no glucose

Operons are found in both prokaryotes and eukaryotes, although many eukaryotic structural genes are controlled differently and transcribed individually, rather than in groups.

Gene control in eukaryotes

Transcriptional level

Transcription factors are proteins that determine whether transcription of particular genes occurs or not. They interact with other control mechanisms in cells to regulate the cell cycle and the responses of cells to hormones, other extracellular signals and intracellular changes. These factors bind to promoter and enhancer regions of DNA that are near to each structural gene. RNA polymerase binds to the complex of proteins

Knowledge check 3

Explain how the *lac* operon is controlled when (a) there is no lactose present and (b) lactose is present.

and starts transcription, as shown in Figure 5. Some transcription factors inhibit transcription by preventing RNA binding to DNA. Whether a gene is transcribed or not often depends on the balance between positive and negative controls.

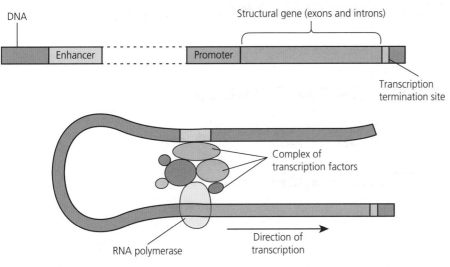

Figure 5 Transcription factors allow RNA polymerase to bind to transcribe a gene; each complex involves many more factors than shown here, including many that bind to other factors

Post-transcriptional level

The mRNA that is produced during transcription is a complete transcript of the gene, including the **introns**, which are non-coding sections between the **exons** — the coding sections. After transcription this **primary mRNA** is edited by having the introns cut out and the exons spliced together. The **mature mRNA** produced then leaves the nucleus through nuclear pores, to be translated.

Not all mRNA is translated. Present in the cytoplasm are microRNA molecules that are about 22 nucleotides in length. MicroRNAs bind to mRNA molecules because they have base sequences that are complementary to sections of mRNA molecules that are not translated. This binding identifies the mRNA for breakdown by enzymes in the cytoplasm.

Exons are not always spliced together in the same sequence. Often they are spliced together in different ways to give different polypeptides coded by the same gene, for expression in different tissues.

Exon Part of a eukaryotic gene that codes for part of a polypeptide. Exons are separated from each other by introns.

Intron Part of a eukaryotic gene that separates exons and does not code for any part of a polypeptide.

Primary mRNA The mRNA that is produced by transcription and includes exons and introns.

Mature RNA The mRNA after introns are cut out and exons are spliced together.

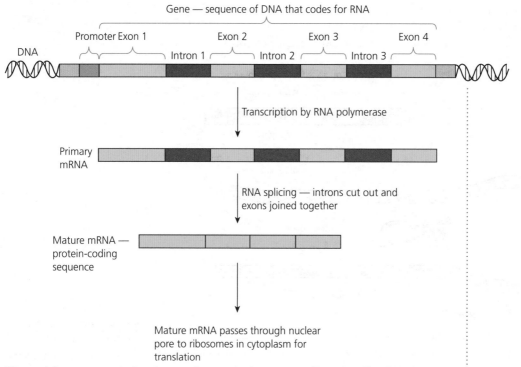

Figure 6 Post-transcriptional control — producing mature (functional) mRNA from primary mRNA

Post-translational level

Enzymes that already exist in the cytoplasm can be activated and inactivated by compounds such as cyclic AMP (cAMP). cAMP acts as a second messenger to activate a protein kinase enzyme at the start of a cascade in response to adrenaline and glucagon. cAMP fits into an allosteric site on the enzyme and changes its tertiary structure. The enzyme becomes active (Figure 7), catalysing the phosphorylation of another enzyme in the cascade (see page 39 in the third guide in this series).

Knowledge check 4

Suggest the advantages of the control of gene expression at each of these three levels.

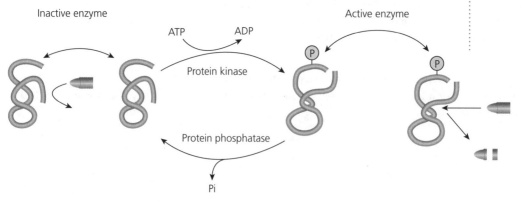

Figure 7 Post-translational control

Mutation

Proofreading by DNA polymerase during replication means that mistakes are rare events and base sequences of genes remain the same generation after generation. However, mistakes do happen and these lead to mutations. Gene mutations involve relatively small changes to the sequence of nucleotides in DNA. Mutations happen spontaneously, but mutagens, such as radiation (e.g. X-rays) and certain chemicals (e.g. benzopyrene in tobacco smoke), increase the rate of mutation.

Types of gene mutation

In a **substitution mutation** there is a change to one base pair, for example from A–T to C–G. If this happens in the first base pair of a triplet then it is likely to change the amino acid that occupies the relevant position in the primary structure of a polypeptide. Any change like this in the primary structure is a **missense mutation**. A change to the second or third base is less likely to change the amino acid and is known as a **silent mutation**.

An **insertion mutation** involves the addition of an extra base pair to a region of DNA. This alters the sequence of bases 'downstream' of the place where the extra base pair was added. A pair of bases is removed from a region of DNA when a **deletion mutation** occurs.

An insertion or deletion leads to a **frameshift mutation**. If a base pair is lost or added, then the 'reading frame' changes during translation so the amino acid sequence downstream of the mutation changes (Figure 8). Depending on where in the gene this happens it can have catastrophic effects on the polypeptide so that it does not work, or works only poorly. Many recessive mutations that do not code for functioning proteins are frameshifts.

Coding   ...A T G T A C G G C T T A C G T T A G...

Template   ...T A C A T G C C G A A T G C A A T C...

mRNA   A U G U A C G G C U U A C G U U A G...
Amino acid reading frames

| met | – | tyr | – | gly | – | leu | – | arg | stop |
| 1 | | 2 | | 3 | | 4 | | 5 | 6 |

Coding   A T G T C G G C T T A C G T T A G...

Template   T A C A G C C G A A T G C A A T C...

mRNA   A U G U C G G C U U A C G U U A G...
Amino acid reading frames

| met | – | ser | – | ala | – | tyr | – | val | – | ser/arg |
| 1 | | 2 | | 3 | | 4 | | 5 | | 6 |

Figure 8 The effects of a frameshift mutation on a sequence of five amino acids; the fifth base pair has been deleted

Some frameshift mutations lead to the formation of a STOP codon where there was not one before. If the mutated gene is transcribed and translated, the polypeptide produced will be much shorter and non-functional. Mutations like this are **nonsense mutations**.

Gene mutation A change to the nucleotide sequence of a gene, giving a mutant allele of the gene.

Exam tip

Look carefully at Table 1 on p. 7 to work out the effects of substitutions of the first, second and third bases in different codons. Try glycine (Gly) to start with.

Knowledge check 5

Explain why the deletion of a triplet of base pairs does not lead to a frameshift mutation.

Knowledge check 6

Summarise the effects of the deletion of the fifth base pair as shown in Figure 8.

Knowledge check 7

What are the effects of an insertion of the base pair A–T after the first triplet in the DNA in Figure 8? (A is on the coding polynucleotide.)

Neutral mutations have no effect on the fitness or survival of the organism. They are of different types:

- The mutant triplet codes for the same amino acid, as happens when CCG changes to CCT (silent mutation).
- The triplet codes for a different amino acid, but this makes no difference to the function of the polypeptide.
- The mutant polypeptide functions in a different way but this does not give an advantage or a disadvantage to the organism.
- The mutation occurs in non-coding regions of DNA and is not expressed in the phenotype.

Most mutations in structural genes that are not neutral are harmful because they disrupt or destroy proteins. There are many examples of these, for example failure to produce chlorophyll in plants (p. 17) and the growth of body parts in the wrong place (p. 15). Other mutations might have little or no effect on the phenotype and so confer no advantage. However, environments can change and selection acts on phenotypes not genes, so a feature that is neutral or disadvantageous can become advantageous when the environment changes or organisms colonise a new habitat. This is why genetic variation is so important (p. 30).

There are far fewer mutations that are known to have beneficial effects. This is probably because over evolutionary time (millions of years) all the possible changes to nucleotide sequences have occurred and those that confer advantages have become 'fixed' in genomes and are present as alleles in populations at frequencies that are not maintained simply by mutation continually occurring. The following examples illustrate the difficulty in deciding whether a mutation is beneficial or not:

- Antibiotic resistance mutations in *Neisseria meningitidis*, such as resistance to the antibiotic rifampin in the gene *rpoB* that codes for part of RNA polymerase, is beneficial to the bacteria when exposed to the antibiotic. Often resistance like this comes at a metabolic cost and reduces the success of bacteria when there is no antibiotic present.
- A resistance mutation in the same gene in *Phytophthora infestans* makes the plant pathogen difficult to control with fungicides, so is not beneficial to humans.
- A mutation in a sulfate transporter in plants reduces the uptake of selenate, an ion that becomes toxic if too much is absorbed. This gives plants the ability to colonise certain soils, but is of no advantage in other soils.

Mutations might not always be beneficial to every individual. Many are beneficial when in heterozygotes. The best-known example is the substitution mutation in the sixth codon of the first exon of the gene for β-globin in haemoglobin. This changes glutamic acid to valine, which alters the functioning of haemoglobin. People who are heterozygous have a reduced risk of developing malaria. People who are homozygous for the mutant allele develop sickle-cell anaemia (Table 3 on p. 21).

Genes in development

Some transcription factors activate genes, so controlling the sequence of developmental changes that occur when a zygote becomes an embryo. Cells differentiate and become specialised for specific roles. To do this, they switch some genes on and some off in a carefully organised sequence.

Exam tip

One of the most famous beneficial mutations occurred in the peppered moth, *Biston betularia*, resulting in black coloration (p. 30).

Knowledge check 8

There are genes that code for tRNA molecules. What might be the effect of a mutation in the DNA that codes for the anticodon region of a tRNA molecule?

Homeotic genes control the development of the structures that develop from each segment in the head and thorax of insects. Mutations of these genes lead to adult flies with grotesque phenotypes, with body structures, such as legs, antennae and wings, growing in the wrong places — a phenomenon known as homeosis. These genes code for transcription factors that bind to DNA.

The homeobox is a sequence of bases in DNA that codes for the homeodomain, which is a region of 60 amino acids with a specific shape that binds to DNA to regulate transcription. A domain is a region of a globular protein that carries out a specific function. The homeodomain is composed of three α helices that bind to the grooves of the DNA double helix. Some of these genes are Hox genes, which control the development of each segment of the body. The homeobox sequences of organisms are almost identical because they all have the same function — they code for DNA-binding transcription factors. They have been conserved during evolution.

The eight Hox genes in *Drosophila melanogaster* are in two clusters on chromosome 3 and are positioned along the chromosome in the same order as the body segments in which they are expressed.

> **Exam tip**
>
> Search for information about antennapedia in *Drosophila*. This is one of the mutations that prompted the quest that led to the discovery of the Hox genes.

The role of genes in development

Mitosis, cytokinesis and apoptosis are involved in growth and development. New cells are formed by mitosis and cytokinesis and these become specialised to form different organs and tissues. The cell cycle is controlled by genes, most of which code for transcription factors that regulate gene expression so that the proteins needed for each stage of the cell cycle are produced in the correct sequence. These genes also respond to signalling compounds released by other cells to increase their rates of mitosis for growth, reproduction and replacement of dead cells.

Apoptosis is programmed cell death, which happens throughout life, but is an important part of development. Cells respond to external and internal signals that trigger an ordered sequence of changes in the cytoplasm, ensuring that cells are removed efficiently without the release of hydrolytic enzymes that would damage surrounding tissue and cause inflammation (Figure 9).

Transcription factors respond to signals that stimulate the production of proteins that promote apoptosis. These activate caspases, which are part of an enzyme cascade that breaks down structural proteins such as those of the cytoskeleton. Anti-apoptosis genes code for inhibitor proteins that prevent apoptosis in healthy cells. Cells between the developing fingers and toes are destroyed so that they become separate structures. It also occurs during the formation of synaptic connections in the nervous system. Apoptosis can also be a response to stress, such as infection in plants, because it is a way to destroy cells infected with viruses.

Homeotic gene Any gene that codes for a transcription factor that regulates development of parts of the body in animals, plants and fungi.

Homeobox The nucleotide sequence that codes for the homeodomain.

Homeodomain The part of a transcription factor coded by homeotic genes, and which binds to DNA.

Hox gene A homeobox gene that controls the development of features in each segment of the body.

Knowledge check 9

Explain why homeobox genes in different organisms have almost identical base sequences.

Apoptosis Programmed cell death, which is controlled by specific proteins so that harmful products are not released into the body. It is an essential part of development.

1

Cell destined to go through apoptosis. Proteases break down cell proteins, especially those of the cytoskeleton

2

Cell shrinks as cytoskeleton is broken down and organelles become tightly packed together. Plasma membrane pulls away adjacent cells

3

DNA in the nucleus becomes denser and tightly packed, forming darkly staining areas of chromatin just inside the nuclear envelope

4

Nucleus breaks up into smaller pieces

5

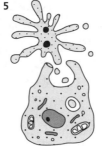

Small, membrane-bound vesicles of cytoplasm 'bud' off from the surface of the cell — a process known as 'blebbing'. Vesicles are engulfed and destroyed by phagocytes, so cell contents are not released into the body

Figure 9 The events that occur during apoptosis — programmed cell death

Synoptic links

This is a good opportunity to revise the structure of prokaryotic and eukaryotic cells from Module 2. You could be asked where the stages in protein synthesis occur in eukaryotic cells. The role of cAMP as a second messenger is covered in Module 5. The cell cycle is covered in Module 2.

Summary

- Genes are expressed in cells when they are transcribed and translated to form a protein that becomes active. Gene expression is controlled at different stages in protein synthesis.
- The *lac* operon is responsible for the control of the production of three proteins in the prokaryote *Escherichia coli*. Regulation is at the level of transcription.
- Gene regulation in eukaryotes operates at three different levels: transcriptional, post-transcriptional and post-translational.
- Transcription factors are proteins that control whether transcription happens or not. The molecule of mRNA formed immediately after transcription in eukaryotes is primary mRNA. This is edited by the removal of sections known as introns. The remaining sections of RNA are exons and these are joined together to form mature RNA, which leaves the nucleus for ribosomes in the cytoplasm.
- The binding of cyclic AMP to proteins is one way in which post-translational control occurs. This binding may activate proteins, as is the case with protein kinases.

- Gene mutations change the sequence of nucleotides in DNA molecules. Substitution mutations result in a change to a single amino acid in a polypeptide. Deletion and insertion mutations result in a change to the sequence of amino acids downstream from the site of the mutation. This effect on the protein is known as a frameshift.
- Most mutations have a harmful effect on the phenotype; some substitution mutations can have a beneficial effect on the phenotype. Neutral mutations have no effect on the phenotype.
- The genes that control the development of body plans are similar in plants, animals and fungi, and are highly conserved. The homeobox sequences are sequences of nucleotides in these genes that code for transcription factors that bind to DNA. Their sequences are very similar because they code for the domain that binds to DNA.
- Cell division by mitosis is responsible for growth in eukaryotic organisms. It is responsible for the production of new cells, which differentiate into all the cell types.
- Apoptosis is programmed cell death that changes body plans, for example by removing tissue between digits (fingers and toes). It is also involved in the response to stress, such as infections.

■ Patterns of inheritance

Key concepts you must understand

Variation refers to differences between the phenotypes and genotypes of organisms. Phenotypic variation is obvious to us in the physical appearance of organisms and in their behaviour. It is also evident in aspects of biology that we cannot see so easily, such as biochemistry and physiology.

Genetic variation is detectable by using techniques such as DNA sequencing to investigate the types of gene and their variants (**alleles**) present in different organisms. Examples of this are the variations in haemoglobin (p. 20) and blood group antigens (p. 21).

Key facts you must know

Genotype and environment contribute to phenotype

The genotype is directly responsible for certain phenotypic features. Your blood group is determined by the alleles of the *ABO* gene (Table 2). The environment has no influence at all on the phenotype for this feature. Most of the features that show discontinuous variation are controlled only by genes (p. 24).

Genotype and environmental factors often interact to influence the phenotype — for example, many genes influence the growth of organisms. Whether that genetic capability is achieved or not depends on environmental factors such as diet. Mice have genes that severely limit growth and others that encourage more growth than normal. A mutant allele of the gene *lit* leads to less growth hormone (GH) produced by the anterior pituitary gland, causing dwarfism. Mutations that disrupt the response of cells to GH can lead to gigantism in mice. Mice homozygous for the mutant allele *ob* eat large quantities of food and become obese. The environment also has an effect on body size. Mice without mutations like those described fed on a diet deficient in calcium and/or vitamin D have poor skeletal growth and do not grow to full size.

Light influences growth in seedlings. It reduces the production of auxin so that seedlings grow thick stems that can support leaves. Light stimulates leaves to expand and turn green. In the dark, these changes do not occur, so seedlings show etiolation — they are tall and spindly, with tiny leaves. Light is needed to stimulate the development of chloroplasts and production of chlorophyll, so these seedlings are yellow because carotenes are still produced. The absence of chlorophyll is **chlorosis** and plants without chlorophyll are described as chlorotic.

Some seedlings never produce chlorophyll even when in full Sun. This is because they have inherited two alleles of one of the genes needed for chlorophyll synthesis. These seedlings grow for as long as they are supported by the energy stores in the seed. Plants that are deficient in nitrogen and/or magnesium also show chlorosis, in which the leaves turn blotchy and yellow because they cannot make enough chlorophyll.

Allele A variant of a gene. All the alleles of a gene occupy the same locus (position) on a chromosome and code for the same protein. Alleles have different nucleotide sequences which may differ in only one base pair.

Etiolation The phenotype of plants grown in the dark: tall, spindly stems; small, yellow leaves; and undeveloped chloroplasts.

Chlorosis The phenotype of plants that have no chlorophyll, due to a genetic or an environmental factor.

The importance of sexual reproduction

Sexual reproduction is responsible for generating genetic variation. There are two aspects to this:

- meiosis
- fertilisation

The effects of random segregation and crossing over during meiosis are to 'shuffle' the genes that an individual has inherited before passing them on to the next generation. You should know four ways in which meiosis generates inheritable variation:

- The alleles of one gene segregate, so daughter nuclei contain one of each pair of alleles.
- Homologous chromosomes show random segregation, so daughter nuclei contain a mixture of paternal and maternal chromosomes.
- Homologous chromosomes cross over during prophase I, so daughter nuclei receive chromosomes that are a mixture of paternal and maternal segments.
- The segregation of sister chromatids is also random — often one chromatid is a mixture of maternal and paternal DNA while the other one is not.

The gametes produced by meiosis show a huge range of variation as a result of meiosis. Mating in large populations is usually at random so the genotypes of the two parents are often quite different. The fusion of gametes occurs at random, so this adds another cause of variation. Mating is not always at random — in small populations random mating is not possible (see genetic drift on p. 30). In some populations assortative mating occurs between individuals with similar phenotypes or behaviours.

Without variation there can be no selection. Variation provides the 'raw material' for selection.

Monogenic inheritance

A genetic diagram, such as Figure 10, shows the pattern of inheritance of one gene with two alleles. This is an example of monogenic inheritance. The diagram shows the inheritance of the gene for wing length in *D. melanogaster*. There are two alleles: long wing, **W**, also known as wild type, and vestigial, **w**. Vestigial means very small. **W** is the dominant allele; **w** is the recessive allele. During meiosis in the F_1 flies the alleles separate because they are on homologous chromosomes that separate during meiosis I. This separation is called segregation of allelic pairs. The two alleles of any gene in a diploid cell pass into different gametes during meiosis. Fertilisation gives rise to genetic variation because gametes with the allele **W** can fuse with gametes with **W** or **w** to give flies with the homozygous dominant genotype, **WW**, and flies that are heterozygous, **Ww**. These flies all have long wings, so dominance reduces variation in populations. Compare this with codominance below.

Exam tip

This is a good point at which to revise the details of meiosis, which is in Module 2 (pp. 63–67 of the first guide in this series).

Monogenic (monohybrid) inheritance The transmission of a characteristic controlled by a single gene from generation to generation.

Segregation The separation of the alleles of a gene (allelic pair) during anaphase I of meiosis.

Dominant allele An allele that determines the phenotype when in a heterozygous genotype, for example **Ww**.

Recessive allele An allele that determines the phenotype only when in a homozygous genotype, for example **ww**.

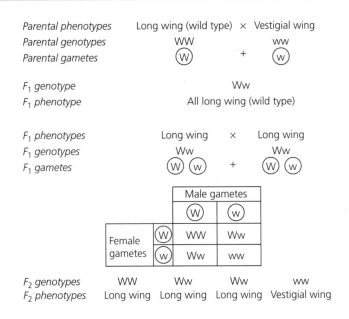

	Parental phenotypes	Long wing (wild type) × Vestigial wing
	Parental genotypes	WW ww
	Parental gametes	Ⓦ + Ⓦ

F_1 genotype   Ww
F_1 phenotype   All long wing (wild type)

F_1 phenotypes   Long wing × Long wing
F_1 genotypes   Ww   Ww
F_1 gametes   Ⓦ Ⓦ + Ⓦ Ⓦ

	Male gametes	
	Ⓦ	Ⓦ
Female gametes Ⓦ	WW	Ww
Ⓦ	Ww	ww

F_2 genotypes   WW   Ww   Ww   ww
F_2 phenotypes   Long wing Long wing Long wing Vestigial wing

F_2 phenotypic ratio   3 long wing : 1 vestigial wing

Figure 10 A monohybrid cross: the inheritance of the gene for wing length in *Drosophila melanogaster*

The Punnett square (Figure 10) is the best way to show the genotypes of the next generation, even when there are few genotypes involved. Note that the Punnett square shows all the possible outcomes — the genotypes do not represent actual organisms. It also shows the probabilities of different genotypes and phenotypes in the next generation.

Test crosses are used to determine genotypes. For example a long-winged fruit fly may be homozygous dominant or heterozygous. It is important to know the genotype of individual organisms used in breeding experiments and in selective breeding. Individuals with unknown genotypes are crossed with individuals that are homozygous recessive, so all the alleles of the former will be expressed in the test cross offspring.

Codominance

Sometimes there is no dominance between two (or more) alleles at the same locus. Both alleles are expressed in the phenotype of a heterozygote. The M and N blood groups show this **codominance**. Two alleles code for glycoprotein antigens (M and N) on the surface of red blood cells. The genotypes and phenotypes are as follows:

- $GYPA^M GYPA^M$ — glycoprotein M; blood group M
- $GYPA^M GYPA^N$ — glycoproteins M and N; blood group MN
- $GYPA^N GYPA^N$ — glycoprotein N; blood group N

Other examples of codominance are coat colour in cattle (red, white and roan), flower colour in the four o'clock plant, *Mirabilis jalapa*, and in snapdragons, *Antirrhinum majus*, and the ABO blood group system in primates. These are all examples of discontinuous variation. Codominance increases variation compared with features controlled by genes that show dominance.

Knowledge check 10

Meiosis and fertilisation are two causes of variation within a species. State another cause of variation.

Exam tip

If you draw Punnett squares in genetic diagrams rather than joining gametes with lines you are less likely to make mistakes in deriving all the possible genotypes.

Exam tip

Test crosses are often asked for in exam questions. Make genetic diagrams to show test crosses for features in fruit flies in both monohybrid and dihybrid crosses (see below).

Codominance Both alleles have an effect on the phenotype of a heterozygous individual.

Multiple alleles

Many genes have more than two alleles – **multiple alleles**. There are three alleles at the human ABO blood group locus, which is on chromosome 9 of the human genome. The gene controls the production of antigens on the red blood cells. They are called antigens because if red cells from one person are injected into an experimental animal they are detected as 'foreign' and stimulate the production of antibodies. If red blood cells are injected into another human then sometimes they stimulate the same response, but not always. This is why blood must be typed before a blood transfusion.

The symbol used for the ABO gene is I and it has three alleles, I^A, I^B and I^o. I^A and I^B are codominant and both are dominant to I^o. Any individual can only have two of these alleles because humans are diploid, with two copies of chromosome 9 in each cell. Table 2 shows the genotypes and phenotypes.

Table 2 Genotypes and phenotypes of the ABO gene

Genotype	Phenotype
$I^A I^A$	Blood group A
$I^A I^o$	
$I^A I^B$	Blood group AB
$I^B I^B$	Blood group B
$I^B I^o$	
$I^o I^o$	Blood group O

The genetic diagram in Figure 11 shows how it is possible for a man who has blood group A and a woman who has blood group B to have children with all four blood groups.

Parental phenotypes   Group A   ×   Group B

Parental genotypes $I^A I^o$ × $I^B I^o$

Gametes I^A I^o × I^B I^o

		Female gametes	
		I^B	I^o
Male gametes	I^A	$I^A I^B$	$I^A I^o$
	I^o	$I^B I^o$	$I^o I^o$

Offspring genotypes $I^A I^B$ $I^A I^o$ $I^B I^o$ $I^o I^o$

Offspring phenotypes (blood groups)   AB   A   B   O

Figure 11 The inheritance of the gene for the ABO blood group system, which has three alleles. The ABO system is an example of codominance because both alleles, I^A and I^B, are expressed to give the blood group AB

The phenotypic ratio is 1:1:1:1. In human genetics it is more usual to give the probability of a child inheriting a particular blood group. In this family the probability that a child will have any one of these blood groups is 0.25 (25% or 1 in 4).

Multiple alleles When a gene has more than two alleles. Any diploid individual can only have two of these alleles in its genotype.

The ABO system and sickle-cell anaemia (SCA) are examples of discontinuous variation (Table 3).

Table 3 Comparison between the ABO blood group system and sickle-cell anaemia in humans

Feature	ABO blood group system	Sickle-cell anaemia (SCA)
Location of gene	Chromosome 9	Chromosome 11
Gene	*ABO* — codes for an enzyme that adds a sugar molecule to a cell-surface antigen	*HBB* — codes for the β-globin polypeptide in haemoglobin
Number of alleles	Three (I^A, I^B, I^O)	Two (**Hb^A**, **Hb^S**)
Number of genotypes	Six	Three
Number of different phenotypes	Four (A, B, AB and O)	Two or three (normal; sickle-cell trait; sickle-cell disease)

Fur colour in rabbits is another example. There are four alleles with a dominance hierarchy:

- **C** is dominant to **c^h**, **c^{ch}** and **c^a**.
- **c^h** is dominant to **c^{ch}** and **c^a**.
- **c^{ch}** is dominant to **c^a**.
- **c^a** is recessive to the other three alleles.

Table 4 shows the genotypes and phenotypes of this multiple allelic system.

Table 4 Genotypes and phenotypes of the coat colour gene

Genotype	Phenotype
CC, Cc^h, Cc^{ch}, Cc^a	Agouti fur
c^{h}c^h, c^{h}c^{ch}, c^{h}c^a	Himalayan fur
c^{ch}c^{ch}, c^{ch}c^a	Chinchilla fur
c^{a}c^a	Albino fur

Sex linkage

One of the genes that controls eye colour in fruit flies is on the X chromosome. Some fruit flies have white eyes because there has been a large deletion in this gene. The alleles are **R** for red eyes and **r** for white eyes. Table 5 shows the five different genotypes and their phenotypes.

Table 5 Genotypes and phenotypes for eye colour in *Drosophila melanogaster*

Males		Females	
Genotype	Phenotype	Genotype	Phenotype
X^{R}Y	Red eyes	X^{R}X^R	Red eyes
X^{r}Y	White eyes	X^{R}X^r	Red eyes
		X^{r}X^r	White eyes

Note that males only have one allele and this is always expressed in the phenotype. Genotypes like this are described as **hemizygous**. When the mutant allele is recessive, heterozygous females are described as (genetic) **carriers**.

Knowledge check 11

A breeder has many rabbits with all four different types of fur. Explain why test crosses are required to find pure-breeding rabbits.

Knowledge check 12

Justify classifying the alleles **Hb^A** and **Hb^S** as codominant.

The X and Y chromosomes should always be included in genetic diagrams involving **sex linkage** — see Figure 12, which shows the inheritance pattern when a pure-bred (homozygous) white-eyed female is crossed with a red-eyed male. Notice how eye colour 'switches' between the sexes in the F_1 generation and all four combinations of eye colour and sex appear in the F_2 generation.

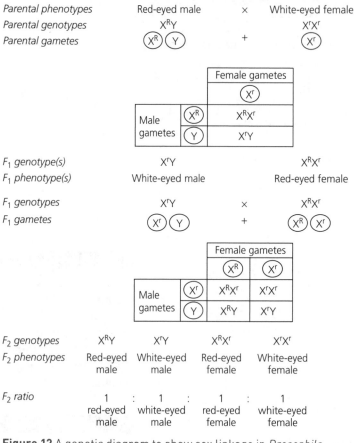

Figure 12 legend details:

Parental phenotypes	Red-eyed male	×	White-eyed female
Parental genotypes	$X^R Y$		$X^r X^r$
Parental gametes	X^R Y	+	X^r

		Female gametes
		X^r
Male gametes	X^R	$X^R X^r$
	Y	$X^r Y$

F_1 genotype(s)	$X^r Y$	$X^R X^r$
F_1 phenotype(s)	White-eyed male	Red-eyed female

F_1 genotypes	$X^r Y$	×	$X^R X^r$
F_1 gametes	X^r Y	+	X^R X^r

		Female gametes	
		X^R	X^r
Male gametes	X^r	$X^R X^r$	$X^r X^r$
	Y	$X^R Y$	$X^r Y$

F_2 genotypes	$X^R Y$	$X^r Y$	$X^R X^r$	$X^r X^r$
F_2 phenotypes	Red-eyed male	White-eyed male	Red-eyed female	White-eyed female
F_2 ratio	1 red-eyed male	: 1 white-eyed male	: 1 red-eyed female	: 1 white-eyed female

Figure 12 A genetic diagram to show sex linkage in *Drosophila melanogaster*

Sex linkage includes the following features:

- The expression of the sex-linked mutant phenotype is more common in males.
- Male offspring cannot inherit the trait from their fathers, but female offspring can.
- Female offspring may be carriers; male offspring can never be carriers.
- Males inherit their Y chromosome from their father. This chromosome does not have some loci that are found on the X chromosome.

Knowledge check 13

What results would you expect in the F_1 and F_2 if the parents were a white-eyed male and a homozygous red-eyed female?

Exam tip

Prepare for questions on genetics by making genetic diagrams to show inheritance patterns for codominance, multiple alleles and sex linkage. Set out your diagrams as in Figure 10.

Knowledge check 14

Haemophilia and colour-blindness are X-linked recessive conditions. A grandfather has both conditions. Three of his grandsons have both conditions, but a fourth is colour-blind only. No-one in the rest of the family has these conditions. Explain how this happened.

Dihybrid inheritance

The transmission of two characteristics each controlled by a different gene is known as **dihybrid inheritance**. The two genes that control these characteristics may be on the same chromosome, in which case they are described as linked genes. This form of linkage is explained on pages 27–28.

Figure 13 shows what happens in the inheritance of two **unlinked genes** that are not on the same chromosome. Figure 14 shows the arrangement of maternal and paternal chromosomes at metaphase in meiosis I, and how this is responsible for the production of four types of gamete. There is a 50% chance that a cell undergoing meiosis will be like **A** in Figure 14 or like **B**. This gives a 25% chance of forming each of the four gametes with different genotypes. As a result of the random assortment of homologous chromosomes at the equator during metaphase I, the allelic pairs, **W/w** (wing length) and **E/e** (body colour), segregate independently of one another and the gametes receive a mixture of maternal and paternal chromosomes.

> **Dihybrid inheritance**
> The inheritance of two genes that may or may not be on the same chromosome.

Parental phenotypes	Long wing, grey body (wild type) ×		Vestigial wing, ebony body
Parental genotypes	WWEE		wwee
Parental gametes	(WE)	+	(we)

F₁ genotype	WwEe
F₁ phenotype	All long wing, grey body (wild type)

F₁ phenotypes	Long wing, grey body	×	Long wing, grey body
F₁ genotypes	WwEe		WwEe
F₁ gametes	(WE) (We) (wE) (we)	+	(WE) (We) (wE) (we)

		Male gametes			
		(WE)	(We)	(wE)	(we)
	(WE)	WWEE	WWEe	WwEE	WwEe
Female gametes	(We)	WWEe	WWee	WwEe	Wwee
	(wE)	WwEE	WwEe	wwEE	wwEe
	(we)	WwEe	Wwee	wwEe	wwee

F₂ genotypes	W_E_	W_ee	wwE_	wwee
F₂ phenotypes	Long wing grey body	Long wing ebony body	Vestigial wing grey body	Vestigial wing ebony body
F₂ phenotypic ratio	9 : long wing grey body	3 : long wing ebony body	3 : vestigial wing grey body	1 vestigial wing ebony body

Figure 13 A dihybrid cross involving two unlinked genes on different chromosomes

Exam tip

The dash (–) in Figure 13 represents either allele for each gene locus, **W/w** and **E/e**. It is useful to use dashes when completing genetic diagrams; if you do this in an exam answer you should explain what it means.

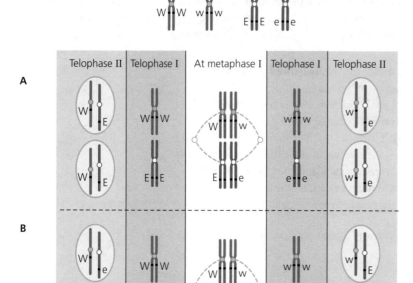

Figure 14 A and B show the two arrangements of maternal and paternal chromosomes in metaphase I responsible for independent assortment during meiosis in the formation of gametes in the F_1 generation

Independent assortment can generate 2^{23} different combinations of maternal and paternal chromosomes (8 388 608). At fertilisation the number of possible combinations is therefore $2^{23} \times 2^{23}$ in a human zygote.

Continuous variation and discontinuous variation

Continuous variation refers to quantitative differences within a species. Examples are shown in Table 6. Each of these examples has no clear categories; instead, there is a range between two extremes — shortest and tallest, lightest and heaviest, etc. Features that show this type of variation are often influenced both by many genes and also by the environment.

Discontinuous variation is the type of variation in which there are clear, non-overlapping categories. This refers to qualitative differences that are in contrasting categories. This type of variation is caused by genes — the environment has no effect. Our ability to taste a bitter chemical, PTC, is determined by a gene (*TAS2R38*) on chromosome 7. The dominant allele gives people the ability to taste PTC; people who are homozygous recessive cannot taste it. The types of food we eat may modify how we taste PTC, but cannot make non-tasters into tasters.

Independent (random) assortment The segregation of allelic pairs at one locus occurring independently of the segregation of another pair at anaphase I of meiosis.

Table 6 Comparison between continuous and discontinuous variation

Feature	Continuous variation	Discontinuous variation
Appearance of phenotype	Quantitative, for example mass and length, with many intermediates	Qualitative, for example presence or absence of a feature, with no intermediates
Number of genes	Many (polygenic)	Few, often just one gene (monogenic)
Effects of alleles at each locus	Small effects	Large effects
Effect of the environment	Large	Small or non-existent
Representation	Frequency histogram	Bar chart
Examples		
Microorganisms	Hyphal length in filamentous fungi, (e.g. *Penicillium chrysogenum*)	Antibiotic-resistant and susceptible strains of bacteria
Plants	Mass, length and width of leaves; height of shoots, length of roots	Tall and dwarf pea plants; flower colour; position of flowers on plant
Animals	Mass, height, length of animals	ABO, Rhesus and MN blood groups in primates; attached and free earlobes in humans

The grain produced by wheat is white or various shades of red. In some varieties the red colour is determined by one gene with two alleles. In others it is determined by two genes, each with two alleles. The effects of these genes and their alleles are additive, as you can see in Figure 15. In other varieties the same feature is controlled by three genes and it is much harder to distinguish between the different shades of red, so the variation is more like continuous than discontinuous.

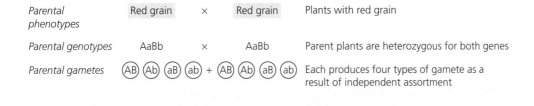

The cross layout:

Parental phenotypes	Red grain	×	Red grain	Plants with red grain
Parental genotypes	AaBb	×	AaBb	Parent plants are heterozygous for both genes
Parental gametes	(AB) (Ab) (aB) (ab) + (AB) (Ab) (aB) (ab)			Each produces four types of gamete as a result of independent assortment

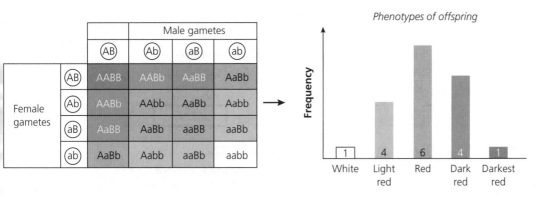

Figure 15 A cross between two wheat plants with red grains of the genotype **AaBb**. The bar chart shows the proportions of the different colours in the phenotypes of the F_2. This is an example of discontinuous variation

Content Guidance

The chi-squared test

The **chi-squared** (χ^2) test is a statistical test used to analyse nominal data. Results are expressed as counts in different categories. The test is used to see if there is a significant difference between the results predicted by the hypothesis being tested and the actual results. The test gives the probability that the difference is due to chance or not.

Fruit flies of the F_1 generation from the cross in Figure 13 were crossed among themselves. The numbers of each phenotype in the F_2 were:

long wings, grey body	134
vestigial wings, grey body	56
long wings, ebony body	60
vestigial wings, ebony body	23

The ratio of phenotypes expected in the test cross offspring of a dihybrid cross such as this is 9:3:3:1, if we assume that the genes are not linked and independent assortment in the F_1 flies has occurred.

The χ^2 test (Table 7), and the probabilities shown in Table 8, are used to find out if these results differ significantly from the expected results or whether any differences are simply due to chance effects.

The **null hypothesis** states that there is no difference between the observed and expected results. The expected results are based on the theory about the inheritance pattern of the gene(s) concerned. The formula for the χ^2 test is:

$$\chi^2 = \Sigma\frac{(O - E)^2}{E}$$

where Σ = sum of…, O = observed value and E = expected value.

Table 7 Calculation of χ^2

Categories	O	E	$O - E$	$(O - E)^2$	$(O - E)^2/E$
Long wings, grey body	134	153.5625	−19.5625	382.6914	2.4921
Vestigial wings, grey body	56	51.1875	4.8125	23.1602	0.4525
Long wings, ebony body	60	51.1875	8.8125	77.6602	1.5172
Vestigial wings, ebony body	23	17.0625	5.9375	35.2539	2.0662
Totals	273	273		$\chi^2 =$	6.5279

> **Exam tip**
> To avoid rounding errors, the expected values and the calculations should be expressed to 4 decimal places. Round up the value of χ^2 to the number of decimal places in the table of probabilities (see Table 8).

Note that the differences between observed and expected results are squared to remove the negative signs, making all figures positive. Each of these figures is divided by the expected number to take into account the numbers of individuals.

The next step is to calculate the degrees of freedom (df). Imagine sorting through the offspring of the test cross. Once you have identified a fly and put it into one of the categories, how many other categories remain? That number is the degrees of freedom. It is calculated as the number of categories minus 1. In this example, $df = 3$. Read across the table of probabilities in Table 8 at $df = 3$ and find the **critical value**, which is 7.82. This is the value of χ^2 where there is a probability of 0.05 (5%) that the results could have been obtained by chance (5% is an arbitrarily agreed figure used by researchers). Where is 6.53?

Table 8 Table of probabilities for χ^2

Degrees of freedom	Distribution of χ^2							
	← Increasing values of p				Decreasing values of p →			
	Probability, p							
	0.99	0.90	0.50	0.10	0.05	0.02	0.01	0.001
1	0.00016	0.016	0.46	2.71	3.84	5.41	6.64	10.83
2	0.02	0.10	1.39	4.61	5.99	7.82	9.21	13.82
3	**0.12**	**0.58**	**2.37**	**6.25**	**7.82**	**9.84**	**11.35**	**16.27**
4	0.30	1.06	3.36	7.78	9.49	11.67	13.28	18.47
	$p > 0.90$ Result is 'dodgy' = too good!		$p > 0.05$ Result is not significantly different from expected outcome			$p < 0.05$ Result is significantly different from expected outcome	$p < 0.01$ Highly significant	$p < 0.001$ Very highly significant

Remember that > means greater than and < means less than. Ruling lines at $p = 0.05$ (or using shading as in Table 8) helps to find out if the χ^2 value is significant or not. See p. 77 for more about this.

In our example, $p > 0.05$ and < 0.1. The probability of getting this result is therefore between 5% and 10%, which means that the result is due to chance effects such as random fertilisation. The difference is not statistically significant and so the null hypothesis can be accepted. If the value for χ^2 is greater than the critical value then the probability < 0.05 and there is a **significant difference** between the observed and the expected results. If there is a significant difference, the prediction is rejected or refined or the experimental procedure is reviewed to look for errors.

> **Exam tip**
>
> The table of probabilities that is used in examination papers is shown on p. 74.

Linkage

When the terms 'linkage' and 'linked' are used about genes without mention of 'sex', then this means the genes exist on the same autosome (non-sex chromosome). An alternative term for this type of linkage is **autosomal linkage**, to distinguish it from sex linkage. With approximately 14 000 genes and only five different chromosomes (one of which is tiny), there is a high probability that any two genes in *Drosophila* are found on the same chromosome. Any two genes on the same chromosome are linked. It is estimated that humans have up to 30 000 genes on 23 different chromosomes. Chromosome 1 (the largest) has over 4000 genes.

Autosomal linkage Two or more gene loci are located on the same chromosome (not a sex chromosome).

Linkage reduces variation because genes are inherited together in linkage groups. Genes that are very close together and always inherited together show **complete linkage**. During **crossing over** in meiosis I, homologous chromosomes pair and exchange portions of non-sister chromatids, as shown in Figure 16. When this happens the genes are described as showing **partial linkage**.

Pairing of bivalent in early prophase I

Chiasma forms between non-sister chromatids in prophase I

Breakage and exchange of parts of non-sister chromatids

Genotypes of gametes = (AB) (Ab) (aB) (ab)

Figure 16 Crossing over during meiosis I between two genes, **A/a** and **B/b**, which are partially linked

The arrangement of linked genes is often abbreviated to $\dfrac{AB}{ab}$ or, in a genetic diagram, to **ABab**, rather than **AaBb**.

If the two gene loci are completely linked, the F_2 offspring will have the genotypes **ABAB**, **ABab** and **abab**, with a 3:1 phenotypic ratio. If they are partially linked then all possible genotypes will appear in the F_2 but not in a 9:3:3:1 ratio. The phenotypes that are different from those of the parents and the F_1 are known as **recombinants**.

Knowledge check 16

Explain the following test cross data:

- long wings, normal bristles   54
- long wings, reduced bristles   7
- vestigial wings, normal bristles   10
- vestigial wings, reduced bristles 63

Epistasis

Epistasis concerns two (or more) genes that influence the same characteristic. Most enzymes work as part of multi-step pathways, such as glycolysis. If an individual is homozygous recessive for an allele that codes for a non-functional enzyme, then none of the enzymes that catalyse reactions later in the pathway can function because they have no substrates. When this happens, the expression of genes that code for enzymes further on in the pathway is suppressed.

The normal (wild-type) eye colour in *D. melanogaster* is a mixture of bright-red pigments and a brown pigment. The genes that determine eye colour code for enzymes in the pathways that produce the pigments and transport proteins that move the pigments into the intracellular structures that hold the pigments. Figure 17 shows the pathway that makes and transports the brown pigment and the phenotypes of flies without functioning enzymes 1 and 3.

Knowledge check 15

Test cross data give the following results: **AaBb** (49), **Aabb** (35), **aaBb** (40) and **aabb** (55). Do the data differ significantly from the expected 1:1:1:1 ratio?

Knowledge check 17

Crossing over does not occur in male *Drosophila*. What effect does this have on variation?

Epistasis The action of one gene in suppressing the action of one or more other genes that control the same feature.

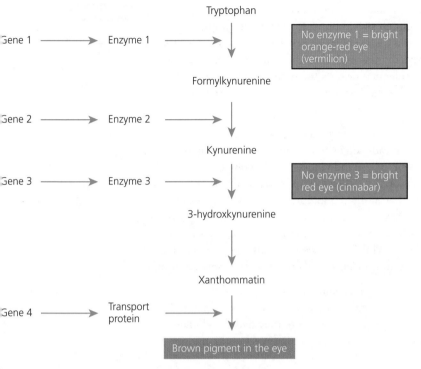

Figure 17 The action of three genes controlling production of the brown pigment that enters the pigmented areas of cells in the eyes of *D. melanogaster*

Fruit flies that are homozygous recessive for gene 1 have vermilion eyes. Even if there is functioning enzyme 2 present, no brown pigment can be produced. Gene 1 is the epistatic gene and gene 2 is the hypostatic gene. If there are two recessive mutant alleles of gene 4 present, no functioning transport protein is produced, so no brown pigment can be moved to mix with the red pigment. Gene 4 is epistatic to all the genes that control this pigment production pathway. The effect of epistasis is to reduce variation.

In some examples of epistasis, one of the gene loci produces an enzyme inhibitor (see Question 7 on p. 73).

Exam tip

You can find good explanations of the common epistatic ratios seen in the F_2 generation by searching for 'Mendel epistasis Philip McLean'.

Factors that affect the evolution of species

In Module 4 you studied Darwin's observations that led him to propose the theory of natural selection. In this section we look at how natural selection acts to stabilise a population and how it can act to change it. Much of Darwin's evidence came from studying aspects of artificial selection, which concentrates on a small number of features, with humans being the selective agent. Sometimes populations change not because of selection but simply as a consequence of being small (genetic drift — p. 30).

Changes to allele frequencies

If an environment stays constant, then selection acts to maintain the population. Selection against extremes in features that show continuous variation maintains the same distribution from generation to generation. The range does not decrease with time because of mutation, meiosis and the influences of the environment. An example of **stabilising selection** is clutch size in birds. If the female parent lays too many eggs it is unlikely that she and her partner can provide enough food, so the young birds could all starve. If she lays too few eggs then there are few descendants to inherit her alleles. Features where selection favours the heterozygotes also lead to stabilising selection; an example is the maintenance of the **Hb^S** allele in human populations in places where malaria is found (Table 3, p. 21).

Changes in abiotic and biotic factors in the environment change the selective pressures that act on organisms. Certain individuals might show features that are adaptations to the changed environment. These individuals have a competitive edge and are able to survive, breed and pass on their alleles. This is **directional selection**. The peppered moth, *Biston betularia*, flies at night and during the day it settles on trees, where it is camouflaged against lichens that grow on bark in unpolluted parts of the British Isles. There are two distinct forms within this species. One has fairly light coloured wings speckled with black. This form is well camouflaged against the lichens that grow on trees. The other form is black and is known as the melanic form.

Before the Industrial Revolution, as soon as any melanic moths appeared they were likely to be eaten by predatory birds because they were so obvious against the speckled background found on trees. They were eaten before they had the chance to reproduce. These melanics reappeared now and again by mutation, but the mutant allele was rarely inherited.

In the middle of the eighteenth century the environment changed significantly. With the Industrial Revolution came severe air pollution from the burning of coal. The smoke contained sulfur dioxide, which killed most of the lichen species, and soot, which was deposited on the trees. In woodlands around Manchester people noticed more and more of the melanic peppered moths, which were now well camouflaged. The speckled moths were easily spotted by birds and eaten while the melanics survived and reproduced. The melanics left more offspring than the speckled variety so that by the beginning of the twentieth century the melanic form made up over 90% of the peppered moth population in woodlands around industrial cities.

Genetic drift and the founder effect

In small populations, changes in allele frequency can occur at random. It may be pure chance as to which individuals survive and breed. This change in allele frequency is known as **genetic drift**.

Small populations that colonise new areas, especially islands or other isolated ecosystems, may have allele frequencies that are not representative of the main population from which they came. This is the **founder effect** and allied to genetic drift is thought to be the cause of some of the allele frequencies seen in human populations that used to be small, dispersed and isolated.

> **Exam tip**
>
> See p. 32 for an explanation of the concepts of gene pool and allele frequency.

> **Exam tip**
>
> Studies on parasitoid wasps that lay eggs inside caterpillars show the same phenomenon. Lay too many eggs and there is less food in the caterpillar to support them all; lay too few and there are few descendants.

Genetic drift Random change in allele frequencies in a small population as a result of non-random mating, *not* selection.

Founder effect When the allele frequencies in small, isolated populations are different from the allele frequencies in the larger population from which they came.

Genetic bottlenecks

A severe reduction in the genetic variation within a species occurs when there is a catastrophic decrease in its population. This has occurred in many species as a result of natural disasters. Humans have caused many **genetic bottlenecks**. The northern elephant seal, *Mirounga angustirostris*, and the European bison, *Bison bonasus*, were both nearly hunted to extinction. Although numbers have recovered, they have very low genetic diversity. Many plant species have been driven to near extinction by habitat destruction and by the activity of plant collectors. Crop plants show much less genetic diversity than the species from which they were domesticated. Soya, *Glycine max*, is a good example.

Figure 18 shows that it takes many years for the intraspecific genetic diversity to recover following a severe population crash, if it ever does. Note that genetic diversity can only recover by incorporation of new alleles and these only emerge by mutation. Few mutations are beneficial and most will have been 'fixed' in the population pre-catastrophe and are lost when the population crashes.

Genetic bottleneck
The effect of a severe decrease in population size on the genetic diversity within a species.

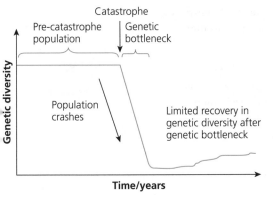

Figure 18 A genetic bottleneck and its effect on genetic diversity

Speciation

The changes described so far apply to changes in populations of a species. According to the definition of the biological species that you used in Module 4, these populations are all members of the same species because they can interbreed and produce viable offspring. With time, a population might change significantly in other ways so that reproduction with other populations becomes impossible. Examples of these isolating mechanisms that separate populations are:

- geographical barriers, such as mountains and seas, prevent individuals mating
- breeding occurring at different times of year
- courtship rituals being different
- hybridisation between two species, producing individuals that are sterile and therefore unable to breed

New species arise from existing species. The process by which this happens is **speciation**. **Allopatric speciation** occurs when populations of a species are physically separated and are in two different geographical areas. **Sympatric speciation** occurs when the two or more populations are in the same geographical area.

Allopatric speciation
Formation of a new species from a population isolated by a geographical barrier.

Sympatric speciation
Formation of a new species within a population as a result of a change that leads to reproductive isolation.

Allopatric speciation

If individuals of a species migrate to occupy a new area, they are exposed to different selection pressures compared with the area that they have left. Over time this may lead to changes in the isolated population to the extent that they cannot interbreed with the original population. Populations of the snapping shrimp, *Alpheus*, were separated when the isthmus of Panama was formed 3 million years ago. Populations on either side of the isthmus are very similar, but when males and females from these two populations are put together they 'snap' at one another and do not mate.

Sympatric speciation

This usually occurs as the result of an abrupt change in a species, so that individuals are not able to interbreed. Individuals of many plant species become **polyploid**, having more than two sets of chromosomes. Often these polyploid types are unable to breed with diploid plants and so become reproductively isolated populations co-existing in the same habitat. Over time, populations evolve differently, but initially it is difficult to tell them apart.

Polyploidy Having more than two sets of chromosomes, for example triploid ($3n$) or tetraploid ($4n$).

Hybridisation sometimes occurs between two different species of plant. The hybrids are sterile and not able to breed with either parent. However, if the chromosome number doubles by a failure of meiosis they become fertile.

Sympatric speciation appears to be less common in animals. Midas cichlids are bottom-dwelling fish that live in a crater lake in Nicaragua. Some time in the past some of these fish colonised the upper, open waters of the lake and bred amongst themselves. They now have a different body shape and are reproductively isolated from the rest of the fish. They are classified as two species — the bottom-dwelling *Amphilophus citrinellus* and the open-water *A. zaliosus*.

The Hardy–Weinberg principle

The Hardy–Weinberg principle states that the frequencies of alleles within a population remain constant from generation to generation unless there is selection in favour of a particular genotype. The principle assumes that:

- the population is large
- mating occurs within a population at random (mating in a small population is non-random)
- there is no mutation
- there is no immigration or emigration
- there are no selection pressures

The allele frequency is used as a measure of the degree of selection that occurs. To understand this you need to appreciate the concept of the **gene pool**. This is all the alleles of all the genes that exist in an interbreeding population. Each individual is diploid and so contributes two alleles (for each gene) to the gene pool.

The Hardy–Weinberg principle is used to calculate the frequencies of alleles. Equation 1 is for alleles; equation 2 is for genotypes.

$p + q = 1$ (Equation 1)

where p is the frequency of the dominant allele, **A** (or whatever allele is being investigated), in the population and q is the frequency of the recessive allele, **a**, in the population.

A modified Punnett square shows how equation 2 is derived. Remember that p and q represent the frequencies of alleles in the gene pool. So this shows the likely outcomes if all individuals can mate at random.

		Frequency of female gametes in the gene pool	
		p (A)	q (a)
Frequency of male gametes in the gene pool	p (A)	p^2 (**AA**)	pq (**Aa**)
	q (a)	pq (**Aa**)	q^2 (**aa**)

$p^2 + 2pq + q^2 = 1$ (Equation 2)

p^2 is the frequency of the genotype **AA**, $2pq$ is the frequency of the genotype **Aa**, and q^2 is the frequency of the genotype **aa**.

(If there are three alleles of the gene then the equations are expanded, thus: $p + q + r = 1$; $p^2 + 2pq + 2pr + 2qr + q^2 + r^2 = 1$.)

If we investigate a population of mice and find that 16% have the recessive phenotype (they have the genotype **aa**) we can say that the frequency of the recessive allele **a** is the square root of 0.16, which is 0.4. This means that the frequency of the dominant allele **A** is 0.6 ($p = 1 - q$). We can then calculate the frequency of the homozygous dominant genotype as **AA** = 0.6^2 = 0.36 (36%) and of the heterozygous genotype as **Aa** = 2 × (0.6 × 0.4) = 0.48 (48%).

We can do this because the homozygous recessive individuals in the population have a recognisable phenotype. We cannot count the number of homozygous dominant individuals to find p^2 because their phenotype is indistinguishable from that of the heterozygotes. We can use the equations to calculate the frequency of **A** in the population and therefore the frequencies of **AA** and **Aa**.

The Hardy–Weinberg equations can be used to see if allele frequencies have changed. If the frequencies of genotypes do not conform to the Hardy–Weinberg principle then one or more of the conditions above does not apply. It may be that selection is occurring.

Knowledge check 19

The frequency of the blood group alleles in a population are: **I^A** = 0.3, **I^B** = 0.1, **I^O** = 0.6. Calculate the frequencies of the three blood groups in the population.

Knowledge check 20

In an adult population in a rural area of Africa affected by malaria the following genotypes are found: **Hb^AHb^A** 605; **Hb^AHb^S** 390; **Hb^SHb^S** 5. Use the Hardy–Weinberg principle to find out if this population is in equilibrium.

Exam tip

If you do an image search for the Hardy–Weinberg equilibrium, you will find graphs that show the effect of changing allele frequencies on genotype frequencies.

Knowledge check 18

Cystic fibrosis is a recessive condition that affects about 1 in 2500 babies in the UK. Use this information to calculate (a) the frequencies of the dominant and recessive alleles, **F** and **f**, in the UK population, (b) the percentage of heterozygous individuals (carriers) in the population.

Artificial selection (selective breeding)

Environmental factors — biotic and/or abiotic — are the agents of selection that operate on populations in natural selection. Humans are the agents of selection in artificial selection because they choose which individuals will survive and breed to pass on their alleles. They often decide the matings that will occur.

Plant breeding

Plant breeders often wish to combine features from different varieties of the same species. Many commercial varieties of crop plant that give good yields at harvest do not have resistance to diseases or pests and cannot tolerate harsh environmental conditions, such as drought. These desirable features may be present in older, cultivated varieties or in wild relatives of the crop plant.

The breeder will cross different varieties by transferring pollen that contains the male gametes. Precautions are taken to make sure that the plants are not pollinated naturally. Often the flowers will have their anthers (that produce pollen) removed and they will be covered in bags to exclude pollen from other plants. Breeders collect the seed, sow it and check the next generation of plants for the desired features. The new hybrid plants are often crossed with commercial varieties to make sure that the new variety will have a good yield. This crossing back to commercial varieties continues for several generations. The breeders will only use the plants that combine the desired qualities, for example disease resistance and high yield. Breeding programmes continue for up to 10 years before new varieties are made commercially available.

Plant breeders improved bread wheat by incorporating alleles from different species. The gene *Sr2*, responsible for resistance to stem rust, a fungal disease, was transferred from *Triticum turgidum* into common (hexaploid) wheat to produce a variety called Hope. It has since been used in other varieties. Mutant alleles that do not code for the enzymes required to make gibberellins have also been incorporated to give dwarf varieties that have increased grain yield.

Animal breeding

Selection is different with domesticated animals, such as cattle, because they are much larger and have a longer life cycle than most crop plants. Breeders choose features of their animals that they want to improve. Animals in the herd or flock are chosen for their appearance or productivity: these are selected and bred together. The breeder checks the offspring to see if they show the desired features and any improvement. The best are selected, bred together and the process is continued with future generations.

For some species, such as cattle, breeders use sperm from males that have superior qualities. By using artificial insemination, sperm from one bull can fertilise eggs in thousands of cows. Eggs from cows that show superior qualities can be harvested, fertilised *in vitro* and the embryos placed in surrogate cows (p. 47). This protects the superior cow from the risks of pregnancy. Much animal breeding is concerned with productivity, for example:

- increasing milk production of cows
- increasing growth rates of pigs and beef cattle
- improving the quality of meat

Knowledge check 21

Explain why it says '... breed to pass on their alleles' rather than '... breed to pass on their genes'.

Exam tip

Do not confuse selective breeding with genetic engineering. Genetic engineering involves moving genes between different organisms, often across species barriers — something that is not possible with selective breeding.

Hybrid The result of a cross between varieties or between species. Most hybrids between species are infertile, but some are fertile and can therefore reproduce. Hybrids between varieties of the same species are fertile.

Exam tip

Gibberellins are a group of plant growth substances (plant hormones).

Breeders who breed animals for show or for the pet trade are concerned with appearance and temperament.

Artificial selection is not restricted to plants and animals. Research scientists use similar methods to improve strains of bacteria and yeasts for industrial processes, such as yoghurt production, brewing and baking.

Maintaining genetic resources

We rely on very few species of crop plant and domesticated animal to provide our food. Most of these species are the result of many years of selective breeding and now have very low genetic diversity. Plant breeders search for alleles that might provide resistance to new pathogens and pests and permit plants and animals to survive in changing conditions, such as drier or wetter environments. This requires the conservation of wild relatives that can interbreed with cultivated varieties. This is done by conserving them in the wild and using botanic gardens, gene banks, seed banks, zoos and 'frozen zoos' as places where whole organisms, gametes, embryos and seeds can be kept for the future.

Ethics of artificial selection

Artificial selection can be taken to extremes. With animals, the best that can be done is to cross an individual showing the desired features with another that is similar. If this continues for generation after generation it leads to severe inbreeding. Genetic faults in animals, such as hip dysplasia in some breeds of dogs, are often the result of inbreeding. Dogs with hip dysplasia are lame in their back legs. The Belgian blue is a breed of cattle that has been bred for very large size. The fetal cattle are so large that in 90% of cases they can only be born by caesarean section, which many consider to be unethical.

> **Synoptic links**
>
> Selection is covered in Module 4. It would be a good idea to revise all the aspects of this and the evidence for evolution (pp. 65–66 in the guide for Modules 3 and 4 in this series). In order to understand the section on cloning it is important to revise mitosis from Module 2 (pp. 57–59 in the guide for Module 2 in this series).

Summary

- Genetic and environmental factors interact to influence many aspects of phenotypic variation, such as height and mass. Chlorosis in plants is caused both by putting them in the dark and by the mutation of a gene involved in producing chlorophyll. Etiolation is a seedling's response to darkness.
- Dominant alleles are always expressed in the phenotype; recessive alleles are only expressed when homozygous, for example **aa**. If two different alleles of one gene are both expressed in the phenotype, they are described as codominant. Some genes have more than two alleles (multiple alleles).
- Sexual reproduction leads to genetic variation within a population as a result of crossing over and independent assortment in meiosis and the random fusion of gametes at fertilisation.
- Linkage is the existence of two or more genes that have their loci on the same chromosome (autosomal linkage). Recombination between linked genes occurs as a result of crossing over. Genes that have their loci on the X or Y chromosome are sex linked.
- Epistasis refers to the interaction between gene loci. There are dominant and recessive forms of epistasis that influence phenotypic ratios in the offspring of genetic crosses.
- The chi-squared (χ^2) test is used to test the significance of the difference between observed and expected results in genetic crosses.

Summary

- Continuous variation is variation in quantitative features, such as length and mass, whereas discontinuous variation is variation in qualitative features, such as colour and shape. Continuous variation is influenced by many genes (polygeny) and the environment; discontinuous variation is influenced by one or a few genes.
- Variation is the raw material for selection — both natural and artificial. Selective agents, such as predators, food supply and climate, can act to stabilise a population so it remains constant from generation to generation or change it so it is adapted to changes in the environment.
- Stabilising selection acts against individuals showing extremes of phenotype so that allele frequencies remain the same over time. Directional selection favours one extreme so that allele frequencies change over time.
- A genetic bottleneck occurs when the population of a species decreases to a very small number, with the loss of many alleles and thus genetic diversity. The founder effect refers to the unrepresentative allele frequencies of a small founder population. Large changes in allele frequency can occur by genetic drift, which is the result of non-random mating within small populations.
- The Hardy–Weinberg principle is used to calculate the frequencies of alleles in populations. These frequencies stay the same from generation to generation if there is no directional selection, mutation, migration or non-random mating.
- Allopatric speciation occurs when populations are separated by a geographical barrier. Sympatric speciation occurs when populations occupying the same area are isolated by reproductive mechanisms.
- Humans choose features to improve in domesticated animals and crop and horticultural plants; they choose the individuals that will breed and then select from among the offspring those that will be used to breed the next generation.
- Conserving old varieties and wild relatives of crop plants and domesticated animals is important to maintain genetic diversity for future challenges, such as new pests and diseases and changes in environmental conditions.
- There are ethical issues surrounding artificial selection, such as creating varieties that carry recessive alleles for severe disabilities.

Manipulating genomes

Key concepts you must understand

The **genome** refers to all the genetic material in an individual or in a species. Remember that all members of the same species have the same genes. Individuals differ genetically in the alleles that they have. Various discoveries in the latter half of the twentieth century made it possible to sequence DNA, to search for specific genes, to cut DNA, amplify it and transfer the products between genomes of different species. The discoveries gave rise to the following:

- **gene sequencing**, using knowledge of DNA replication
- **polymerase chain reaction**, using knowledge of DNA replication and a thermostable enzyme
- using **restriction endonucleases** to cut across both strands of DNA at specific sequences of base pairs (these restriction enzymes protect bacteria by cutting the DNA of viruses that infect them)

- transfer of genes using **vectors**, such as viruses and plasmids
- using **reverse transcriptase** to produce DNA from an RNA template
- separating fragments of DNA by **gel electrophoresis**

Genetic engineering is the term usually applied to the process of genetic modification by means that are not possible using selective breeding.

Key facts you must know

Proteins and fragments of DNA are separated by electrophoresis (Figure 19). This involves putting samples into wells cut into an agarose gel, which is in a tank filled with a buffer solution of an appropriate pH. A direct electric current is applied to the gel and the protein molecules or pieces of DNA move towards an electrode. DNA is negatively charged so moves towards the anode. The distance moved by a fragment depends on size; smaller fragments move further per unit time than larger fragments.

Electrophoresis is similar to chromatography. If you have carried out chromatography on chloroplast pigments you will have seen the coloured pigments separating. DNA is invisible unless a blue or fluorescent dye is added. A radioactive DNA probe (with the isotope ^{32}P) can be used to locate specific sequences. These probes bind to complementary sequences in the DNA, making them show up as dark bands when exposed to X-ray film.

Melt agarose gel in buffer solution

Insert a toothed comb at one end of the tank to make the wells to take DNA samples

Pour in molten agarose gel

Leave gel to set. Place electrodes at either end of the tank

When gel is set pour in buffer solution and remove the comb

Add blue dye to each DNA sample

Add DNA and dye mixture to the wells

Connect electrodes to the power supply

When the blue dye is within 10 mm of the end of the gel, disconnect the power supply

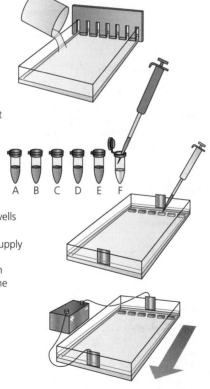

Pour away the buffer and add DNA stain (Azure A) for 4 minutes

Rinse with water and analyse the fragments of the DNA which will appear blue

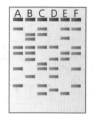

Direction of DNA movement

Figure 19 Electrophoresis to separate fragments of DNA

Content Guidance

Samples of DNA for testing can be very small, for example those retrieved from crime scenes or from chloroplasts and mitochondria. Figure 20 shows how small samples of DNA are amplified in the **polymerase chain reaction**. Primers are short sequences of a polynucleotide that bind to the single-stranded DNA that is being copied. This is necessary for DNA polymerase to start the process of replicating the existing polynucleotide using deoxynucleoside triphosphates (dNTPs). *Taq* polymerase is a DNA polymerase extracted from the thermophilic archaean *Thermus aquaticus*, which lives in hot springs.

Polymerase chain reaction A technique that produces large quantities of specific DNA sequences.

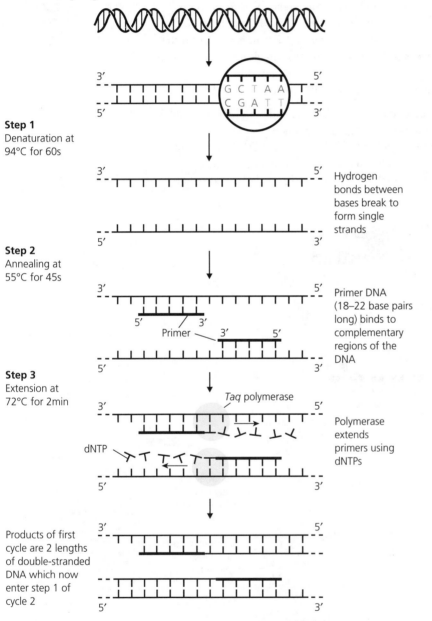

Step 1
Denaturation at 94°C for 60s

Hydrogen bonds between bases break to form single strands

Step 2
Annealing at 55°C for 45s

Primer DNA (18–22 base pairs long) binds to complementary regions of the DNA

Primer

Step 3
Extension at 72°C for 2min

Taq polymerase

dNTP

Polymerase extends primers using dNTPs

Products of first cycle are 2 lengths of double-stranded DNA which now enter step 1 of cycle 2

Figure 20 One cycle of the polymerase chain reaction (PCR). The reaction mixture contains the sample of DNA to be amplified, *Taq* polymerase, Mg^{2+} (cofactor for the enzyme), primer DNA and four different dNTPs

Exam tip

Watch some animations that show the PCR process. Search for the DNA Learning Center and Max Animations.

Exam tip

The Archaea is a domain of organisms that you learned about in Module 4. Extremophiles, such as *T. aquaticus*, are classified in this domain. See p. 60 of the guide to Modules 3 and 4 in this series.

Knowledge check 22

How many lengths of double-stranded DNA are there after eight cycles of PCR?

Knowledge check 23

How many types of dNTP (deoxyribonucleoside triphosphate) are used in PCR?

DNA sequencing

DNA sequencing involves modifying the process of replication to identify each nucleotide as it is added to a growing DNA chain. This enables base sequences of DNA molecules to be read. Current methods of **high-throughput sequencing** generate base sequences of whole genomes in very short periods of time. DNA sequencing is now a routine procedure in many aspects of biology, medicine and forensics.

Genome-wide comparisons are made between the whole genomes of different species to investigate how they may have evolved. Similar comparisons between individuals within a species are made to assess intraspecific variation at the level of base sequences. The effect of differences in gene sequences on phenotypes can also be assessed. For example, the amino acid sequences of proteins are now derived from base sequences using the genetic code (see Table 1 on p. 7). This is much more precise than deriving base sequences from amino acid sequences, which was the procedure before DNA sequencing became highly automated.

Many genes, such as the homeobox sequences of transcription factors (p. 15) and the genes that code for β-globin in different animals (p. 21), are almost identical in very different organisms. Homeobox genes have been conserved through evolutionary time because they code for proteins that bind to DNA and must always have the same tertiary structure and shape. Differences in gene sequences between species show how long ago they diverged (Question 9, p. 82).

> ### Knowledge check 24
>
> Use the genetic code to predict the amino acid sequence from this segment of sequenced DNA. Write out your answer using the one-letter code for amino acids.
>
> 5′ GATGAAGGTGAGAATGAACGCGCGACAGAG 3′

Bioinformatics uses the power of computer systems to hold protein sequences and DNA sequences in databases and make them available for analysis. Researchers might find a DNA sequence in one organism and wish to find other identical sequences. There are programs available that align sequences from different sources and locate identical sequences.

DNA profiling uses sequence data from DNA recovered from crime scenes to help identify perpetrators by comparing their DNA with sequences held on DNA databases or with DNA taken from suspects whose DNA is not already held on databases. These comparisons are made by using base sequences in non-coding regions of DNA.

DNA sequence data are used to identify those at risk of developing medical conditions by searching for the sequences of mutant alleles that are linked to these conditions. They are also used to track the spread of pathogens. This helps identify any differences in the strains of bacteria or viruses that are causing diseases and patterns of transmission. This has proved valuable in the **epidemiology** of the Ebola and Zika viruses.

High-throughput sequencing Any fast method of DNA sequencing that generates sequence data for whole genomes.

DNA profiling Use of a small number of different sequences in DNA to provide genetic identities for people.

Epidemiology The study of the occurrence of diseases, their origins and the factors that influence their spread through populations.

Genetic engineering

All DNA has the same type of structure. This means that pieces of DNA from one organism can be incorporated into the DNA of another (Figure 21). Genes are transferred with promoter sequences so that they will be expressed in the host organism. Other sequences are often incorporated as well to make gene constructs.

Cells that contain 'foreign' DNA — DNA from another organism — translate the code and make the same polypeptide. This sounds quite simple. However, often, when animal and plant genes are inserted into bacteria, although the correct sequence of amino acids is produced, the bacteria cannot cut and fold the polypeptide to produce a functional protein as happens in eukaryotic cells.

The production of new medicines and other chemicals may involve the manipulation of bacteria and yeasts that can make them more easily than other methods. This is an application of **synthetic biology** (see p. 50 in the guide to Modules 3 and 4 in this series).

Synthetic biology
Genetically modifying organisms to synthesise specific drugs and other chemicals.

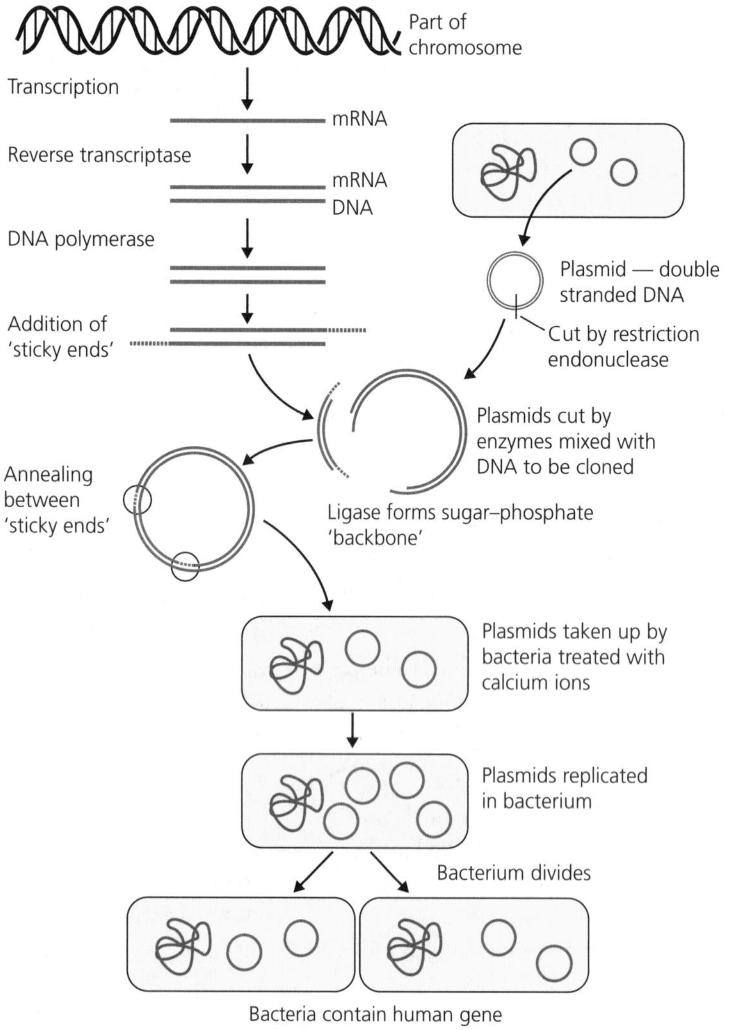

Figure 21 Gene cloning: using bacteria to make multiple copies of a human gene

Stage 1 The gene for cloning is obtained by one of three methods.

1 Restriction endonucleases cut DNA to give a staggered break that leaves 'sticky ends'. (Some cut straight across the DNA to give blunt ends, in which case short sequences of nucleotides are added to give 'sticky ends'.) Different restriction enzymes show specificity by cutting at different restriction sites. The enzyme shown (*Eco*RI) cuts at a site that reads the same in the 5′ to 3′ direction as in the 3′ to 5′ direction, to give 'sticky ends':

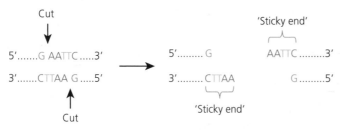

A gene (DNA) probe can be used to check that it is the correct DNA sequence. A gene probe is a small length of single-stranded DNA that can base pair with a base sequence on a longer section of DNA and so locate a specific (complementary) base sequence.

2 RNA is extracted from a cell. The enzyme reverse transcriptase makes a single-stranded copy of DNA, using the RNA as a template. DNA polymerase makes a complementary polynucleotide (cDNA).

3 If the protein sequence is known, then a gene can be manufactured using a 'gene machine' that assembles nucleotides in the desired sequence, using the information from the genetic code.

Stage 2 The gene is inserted into a **vector**.

4 Multiple copies of the gene can be made by PCR.

5 A plasmid is cut open using the same restriction enzyme.

6 'Open' plasmids and the required gene are mixed. Hydrogen bonding between the 'sticky ends' attaches the two — a process known as annealing.

7 The enzyme ligase is added to form covalent phosphodiester bonds to join up the sugar–phosphate backbone of DNA. This is known as ligation.
Recombinant DNA (rDNA) has now been produced (rDNA is DNA formed by combining DNA from two different sources).

Stage 3 Plasmids are taken up by bacteria.

8 Bacteria are treated with an electric field and with calcium ions to increase the chances of plasmids passing through their cell walls. **Electroporation** is used in the **transformation** of animal and plant cells as well.

9 Plasmids enter bacteria, which are now described as transformed because they contain foreign DNA.

10 DNA polymerase in bacteria copies the plasmids; the bacteria divide by binary fission so that each daughter cell has several copies of the plasmid.

11 The bacteria are now **transgenic organisms** because they can transcribe and translate the foreign gene.

Knowledge check 25

Distinguish between these two processes: annealing and ligation.

....

Vector Any means by which DNA is taken into a cell or organism that is to be genetically modified, for example a plasmid or virus.

Recombinant DNA DNA made from DNA from two or more sources, for example a plasmid with a human gene and a bacterial promoter.

Electroporation Entry of foreign DNA into a cell using an electric pulse to increase the permeability of the plasma membrane.

Transformation The process of forming a genetically modified organism.

Transgenic organism Any organism that expresses a foreign gene in its phenotype.

Transformed cells have to be identified because some plasmids do not take up the genes and some bacteria do not take up the plasmids that contain the foreign gene. This can be carried out by inserting genes for antibiotic resistance into the plasmid. Another method is to use a gene for an enzyme that produces a protein that fluoresces in ultraviolet light.

Gene therapy

The transfer of a gene into an organism to repair a genetic fault and thereby treat or cure a genetic disorder is **gene therapy**. There are two forms:

Somatic gene therapy — the gene is inserted into certain cells and cannot be passed on to the next generation.

Germline gene therapy — the gene is inserted into a gamete or into a zygote, so is therefore in every cell of the body. This means that it will be in the cells that give rise to the reproductive cells of that individual — the germline cells — and can be inherited.

In the case of recessive genetic disorders a dominant allele must be inserted into cells where it will be expressed. A good example is the treatment of SCID (severe combined immunodeficiency disease), which is a rare autosomal recessive disease. Children with this condition have no defence against common infections. Some were kept in sterile environments (sometimes known as a 'bubble') in the hope that treatments or a cure would be found.

The non-functioning enzyme is adenosine deaminase (ADA), which is involved in the breakdown of purines, which are toxic to T lymphocytes. In 1990, T cells were removed from a girl who had SCID. They were given the normal allele for ADA, using a retrovirus as the vector, and the transgenic T lymphocytes were placed in her bone marrow. This treatment was successful and has been repeated on several children.

When the genetic disorder is caused by a dominant allele, as it is in Huntington's disease, simply inserting the functional allele will not work because it is masked by the faulty allele. Another problem is that there are many different mutations that give rise to these conditions. A different approach has to be used. One possibility is to use RNA interference (RNAi), in which RNA combines with the mRNA from both the mutant allele and the normal allele. This makes double-stranded RNA, which cannot be translated. This technique effectively silences the genes because it stops the mRNA being translated. DNA for the normal allele is introduced but this is changed slightly so it produces a mRNA that does not combine with the RNAi, and so is translated.

Although gene therapy holds great hope for curing genetic diseases, so far the success rate is low.

Germline therapy is not permitted in the UK and USA. It is illegal in the UK under the terms of the Human Fertilisation and Embryology Act 1990. However, the use of mitochondria from a third party in IVF treatment is permitted in the UK as of 2015. This is used to correct disorders of mitochondrial DNA (mtDNA). There are 150 of these disorders that are passed on from mother to offspring in mtDNA.

Gene therapy The treatment of a genetic disease by replacement of affected cells with genetically modified cells or by introducing functioning genes into affected cells.

Exam tip

There are several different causes of SCID — ADA deficiency being one. Gene therapy has been used successfully to treat an X-linked form of the condition. SCID would make a good basis for an exam question covering aspects of this topic in the context of immunity from Module 4.

Exam tip

Inheritance from mother to offspring is known as maternal inheritance. This would make a good context for a question because it tests knowledge of mitochondria from Modules 2 and 5.

It is unlikely that gene therapy could be used for polygenic conditions, such as coronary heart disease, stroke and diabetes in the near, or even distant, future. In gene therapy of humans, inserted genes may have unforeseen effects, such as disturbing the expression of other genes. If there are such problems, then germline therapy of nuclear genes, if permitted, may have serious long-term consequences for future generations.

Ethics of genetic manipulation

Some of the potential benefits and risks of genetic modification of microorganisms, plants, animals and humans are listed in Table 9.

Table 9 The potential benefits and risks of genetic modification of microorganisms, plants, animals and humans

Genetically modified organisms	Potential benefits	Potential risks
Bacteria	GM bacteria and yeasts produce human proteins that are otherwise in short supply, such as factor 8 (a blood clotting factor), growth hormone and tissue plasminogen activator, which helps to break down blood clots.	GM bacteria and yeasts might escape into the wild and pass their foreign genes to pathogenic microorganisms, making them harder to treat. Antibiotic resistance genes in plasmids are often used as markers to identify transformed bacteria. This could make these antibiotics useless in treating human and animal diseases.
Pathogens for research	Bacteria are modified so they can be grown in the laboratory in order to improve knowledge of their metabolism and help to develop new drugs. Viruses are modified to make suitable vectors in gene therapy.	These GM pathogens could escape from laboratories and compete successfully with non-GM pathogens, causing epidemics.
Crop plants, for example soya	Crop plants with herbicide resistance mean that herbicides can be sprayed during growth of the crop to kill weeds. Crop plants with pest resistance reduce losses to insect pests and reduce the use of insecticides.	Herbicide-resistance genes might be transferred in pollen to species related to the crop. Weeds might become herbicide resistant (sometimes called 'superweeds'). Increases use of herbicides, with subsequent loss of biodiversity.
Mammals for 'pharming'	Transgenic sheep and goats make human proteins in their milk. These proteins are used to treat diseases such as hereditary emphysema.	Insertion of foreign genes might have unforeseen effects.
Gene therapy for humans	Gene therapy has been used to treat various rare genetic diseases, including SCID.	Inserted genes might have unforeseen effects, such as disturbing the expression of other genes, as in the appearance of leukaemia in people treated for SCID. Germline therapy, if permitted, might have serious long-term consequences for future generations if there are such problems.

Aside from the biological risks of genetic engineering, there are ethical concerns over tampering with DNA of different species in ways that could never happen in nature. There are also those who are sceptical of the intentions and procedures of the large biotechnology companies who, they say, are more interested in profits than the long-term welfare of humans and the environment. Many biotechnology companies have patented the genetic modifications that they have developed. To gain an economic return for the years of investment in research and development, they charge farmers a high price for GM seed. Also, farmers cannot keep seed from one year to the next as GM crops do not breed true. The farmers must buy seed each year. This may be difficult for poor, subsistence farmers.

Synoptic links

This topic relies on your knowledge of the structure of DNA and RNA (pp. 30–32 in the guide to Module 2), insulin (pp. 41–44 in the guide to Module 5) and animal defences against pathogens (pp. 42–49 in the guide to Modules 3 and 4). To prepare for questions on ethical issues surrounding cloning and genetic engineering read the genuine concerns of various groups and the counter-arguments from scientists and others. Good places to start are the BioEthics Education Project based at Bristol University and LearnGenetics at the University of Utah.

Summary

- Electrophoresis is the separation of negatively charged DNA fragments in a gel using an electric field. They separate by size. Proteins can also be separated by charge and by size.
- The polymerase chain reaction (PCR) uses a heat-stable enzyme, *Taq* polymerase, to make multiple copies of DNA fragments. This increases the quantity of DNA available for analysis.
- DNA sequencing involves determining the sequence of nucleotides. High-throughput sequencing allows comparisons between the genomes of individuals of the same and different species.
- DNA profiling is used in forensics, epidemiology and assessing disease risk.
- Recombinant DNA is composed of DNA from two or more different organisms, typically from different species.
- Genetic engineering involves extracting genes from one organism, or the manufacture of genes, in order to place them into another organism (often of a different species) so that the receiving (transgenic) organism produces the polypeptide(s) coded by the gene or genes.
- Restriction enzymes remove sections of DNA containing a desired gene by cutting across the sugar–phosphate backbone at restriction sites composed of specific sequences of bases.

- The enzyme reverse transcriptase uses mRNA isolated from cells as a template to produce cDNA.
- DNA probes have base sequences complementary to desired DNA fragments; they are often made radioactive to locate them.
- Plasmids, viruses and liposomes are used as vectors to insert DNA into organisms. DNA fragments are placed into plasmids using the enzyme ligase to form phosphodiester bonds between deoxyribose and phosphate.
- Plasmids are taken up by bacterial cells in order to produce a transgenic microorganism that can express a desired gene product. Calcium ions and electroporation are used to increase the uptake of plasmids by bacteria.
- Somatic gene therapy is the transfer of functioning alleles into cells to treat genetic conditions. Germline gene therapy involves inserting alleles into gametes or zygotes, for the same reason.
- There are many ethical concerns raised by the genetic manipulation of organisms, including the use of microorganisms to produce medicines and other chemicals, gene therapy in humans, manipulating pathogens for research and introducing herbicide and pest resistance into crop plants.

Cloning and biotechnology

Cloning

Key concepts you must understand

Some organisms reproduce asexually so it is possible to perpetuate a strain that shows desirable features. For example, strawberry plants produce runners, which take root and separate from the parent plants. Many crop plants, including all cereals, do not do this and few animals reproduce asexually. Thanks to meiosis and fertilisation the offspring of a prize plant or animal have a different phenotype and most likely not one that is as good or useful. The answer is cloning.

Reproductive cloning produces new individuals with a genotype that is identical to that of the parent individual. Do not confuse this with **non-reproductive cloning**, which is carried out to make cells to treat animals and humans. These are usually formed by cloning an individual's stem cells.

Key facts you must know

Cloning plants

Many flowering plants reproduce asexually by **vegetative reproduction** or vegetative propagation because it involves producing new individual plants by modification of vegetative growth rather than growth from reproduction involving flowers. Propagation is a word meaning multiplication or spreading. Plants can form a clone of genetically identical individuals by growth from leaves, roots and stems. These are then used in horticulture to ensure crops of the same uniform quality.

Cuttings are taken from roots, stems and leaves. Small pieces of a plant are removed and placed into a suitable growth medium. Examples are root cuttings of *Papaver orientale* (oriental poppy), stem cuttings of *Pelargonium zonale* (often mistakenly called 'geranium') and leaf cuttings of *Begonia*.

Tissue culture is a way of producing plants that do not reproduce naturally by asexual means or are hybrids that do not breed true. It is a technique that is useful for ornamental and rare plants, such as orchids. Parts of a plant showing desirable features are removed and used to produce more individuals, as shown in Figure 22.

Meristem culture Meristems contain unspecialised cells that divide by mitosis. Viruses tend not to infect meristems, so meristems can be removed from plants and used to produce virus-free material for cloning in a process called micropropagation. Meristems grow into plantlets that have buds with more meristems, which are subdivided to repeat the process.

Callus culture A callus is a mass of undifferentiated tissue that forms when pieces of shoot, root or leaf are placed in tissue culture. Callus cells can be maintained in tissue culture and subdivided to give large quantities of tissue for cloning. If the growth medium contains appropriate concentrations of plant growth substances (auxins and cytokinins), roots or shoots will develop to give genetically identical plantlets.

Cloning The process of producing a clone, for example by taking individual cells and allowing them to divide into individual organisms.

Clone A group of genetically identical organisms or cells derived from a single cell by repeated division.

Vegetative propagation A form of asexual reproduction in which a plant breaks apart into smaller fragments to form a clone.

Tissue culture Use of tissue removed from an animal or a plant and grown in a special medium.

Micropropagation Removal of meristems and culturing meristematic tissue that is free from viruses.

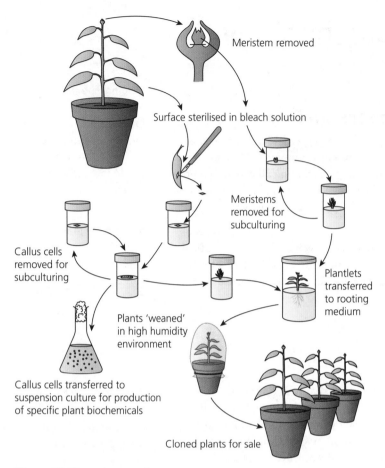

Figure 22 Plant tissue culture

Suspension culture Plant cells from callus cultures are separated from one another and maintained in suspension in a liquid medium. The red dye shikonin is made by plant cells cultured in bioreactors.

Some advantages and disadvantages of cloning plants are listed in Table 10.

Table 10 Advantages and disadvantages of cloning plants in agriculture and horticulture

Advantages	Disadvantages
■ Uniform plants make harvesting easier; the harvested crops show uniform qualities	■ All are susceptible to the same pathogen or pest species, or to changes in climate etc.
■ Makes clones of plants that cannot reproduce sexually to set seed, for example hybrids of lavender or banana plants	■ Propagates single clones that may have genetic diseases and/or do not have resistance to some diseases
■ Can be used to clone transgenic plants	■ Conditions in tissue culture must be kept sterile — aseptic techniques must be used and infected cultures thrown away
■ Allows stocks to be built up quickly; does not have to rely on slower sexual methods of propagation	
■ Tissue culture can be set up anywhere and small plantlets can be transported easily	■ Propagating plants vegetatively is labour intensive and tissue culture requires trained staff and expensive facilities

Cloning animals

Natural cloning occurs by embryo splitting to form identical twins.

Dairy cattle are cloned by using hormones that stimulate the ovaries to produce a large number of eggs. This is known as **superovulation**. These eggs are harvested from the ovary and fertilised *in vitro* by sperm from a superior bull. Each zygote divides by mitosis and the resulting embryo is sexed (to make sure it is XX and not XY), subdivided several times and implanted into surrogate mother cows. This means that the resulting calves are clones of each other, but not of their mother.

This artificial twinning of embryos is a method of cloning that has been used in livestock farming for over 30 years. To clone a high-performing animal this method must be modified, as shown in Figure 23. This method, known as **somatic cell nuclear transfer** (SCNT), is how Dolly the sheep was formed in 1996. Some advantages and disadvantages of SCNT are shown in Table 11.

Somatic cell nuclear transfer Removal of a nucleus from a mature somatic cell, such as a gut cell, and transferring it into an enucleated egg cell.

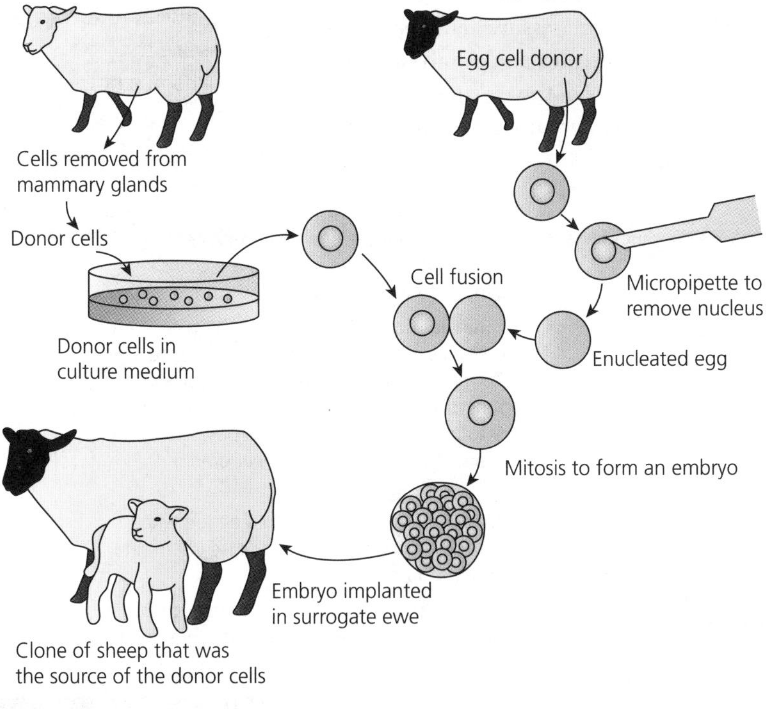

Figure 23 The steps involved in cloning individual sheep and cattle; removal of nuclei from cells is called enucleation

Table 11 Advantages and disadvantages of cloning animals by SCNT

Advantages	Disadvantages
■ Animals giving favourable yields can be produced, for example for high milk yield or quality of meat ■ Prevents rare animals becoming extinct through loss of their genes/alleles ■ Allows fast reproduction of transgenic animals, for example sheep that produce human proteins in their milk	■ Animals may have low quality of life ■ Animals become genetically uniform, which increases susceptibility to disease ■ Success rate has been very low

Biotechnology

Key concepts you must understand

Biotechnology is the use of living organisms in the production of useful products, such as foods and drugs, or in services, such as sewage disposal. Biotechnological processes use either whole microorganisms — bacteria, archaeans, yeasts, mould fungi and protoctists — or enzymes.

Key facts you must know

Culturing microorganisms

A starter culture of bacteria is put into a sterile culture medium in a sterile container. **Aseptic technique** is used throughout to avoid the culture becoming contaminated by competitors and parasites. The culture medium and glassware used are sterilised by placing them in an autoclave, which applies steam under pressure. The top of the sample container is passed through a Bunsen flame that creates an updraft of air so that bacteria from the air do not enter the culture. Samples of the culture are taken at intervals and the number of bacteria counted. The samples are used to estimate the population in numbers per mm^3. Figure 24 shows the results.

Aseptic technique
Procedures followed to ensure that cultures are not contaminated by other microorganisms.

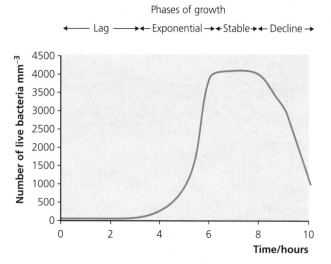

Phases of growth

| ← Lag → | ← Exponential → | ← Stable → | ← Decline → |

- **Lag phase** Growth is slow or non-existent. Reproduction rate is slow as cells are absorbing nutrients, producing enzymes and storing energy.
- **Exponential (log) phase** Reproduction rate is fast. Few cells die. Some bacteria reproduce once every 20–30 minutes, so the population doubles in this amount of time. To fit the numbers onto a graph it is usual to plot the log_{10} (log to the base 10) of the numbers against time (log × linear graph).
- **Stationary phase** A decelerating phase leads into a period when the population remains constant because death rate equals reproduction rate.
- **Decline phase** The death rate is greater than the reproduction rate. The population may decline to zero.

Figure 24 The changes in numbers of bacteria in pure culture; the lag phase to stationary (stable) phase is called sigmoid growth

Knowledge check 27

Draw a graph of your results from Knowledge check 26. To fit the numbers onto a graph plot the log_{10} (log to the base 10) of the numbers against time (log × linear graph). Alternatively plot the numbers directly on to log × linear graph paper. What sort of line do you get?

Knowledge check 26

An imaginary bacterial culture starts with one bacterium. The rate of division is once every 20 minutes. Calculate the numbers of bacteria at every 20-minute interval over 5 hours.

The pattern described in Figure 24 can be explained in terms of the factors that limit growth, which are:

- availability of nutrients
- availability of oxygen
- presence of waste products, some of which may be toxic (e.g. ethanol)

During the lag and exponential phases there are plenty of nutrients and dissolved oxygen. These decrease as the population increases and limit population growth. The accumulation of waste, such as carbon dioxide and ethanol, may make the environment unsuitable, so cells stop reproducing or die.

To find the population of live bacteria, samples are incubated on **agar plates** and the number of colonies is counted. Each colony represents one bacterial cell in the original sample. Alternatively, the sample may be treated with a dye that stains live cells, which are counted under a microscope or with an automatic cell counter.

> **Agar plate** A petri dish with agar that forms a solid medium on which bacteria, yeasts and fungi can grow. Glucose, salts and other nutrients are added to the agar.

Microorganisms are used in biotechnology for the following reasons:

- They grow rapidly (Figure 24) and reproduce asexually to produce genetically identical populations.
- They remain as single cells or small groups of cells, so all cells are productive.
- Among microorganisms there is a wide variety of metabolic processes, with much potential for making different products.
- Strain selection (artificial selection) is easy.
- Genetic modification is easy (Figure 21 on p. 40).
- They carry out complex processes that would be difficult or impossible by chemical means, for example synthesis of fine chemicals such as penicillin.
- It is possible to use cheap sources of nutrients — often waste products from other industries.
- They do not need complex growth conditions.

In some cases the product is a whole organism, as it is with Quorn™ (mycoprotein) production. In most cases the product is a metabolite — a chemical produced during metabolism. **Primary metabolites** are compounds produced during the organism's normal metabolism for its survival, such as amino acids and proteins. Ethanol is an example — it is the waste product of anaerobic respiration in yeast. **Secondary metabolites** are compounds produced that are not essential to the organism's survival. Penicillin is produced by the fungus *Penicillium chrysogenum* when it stops growing. The production of secondary metabolites may provide a way of using enzymes while the organism is not growing.

There are many advantages of using microorganisms in food production:

- Fast growth gives high yields.
- Growth is not dependent on seasons — production can occur throughout the year.
- Waste materials from other industries are used as substrates.
- Factories containing fermenters are not as extensive as farms and can be established where the necessary infrastructure and raw materials exist.
- It is much faster and easier to carry out selective breeding and genetic engineering with microorganisms than with domesticated plants and animals.

- There are fewer ethical issues compared with keeping livestock, although some people object to the use of genetically modified organisms in their food.
- Quorn™ (a meat-like food made from the fungus *Fusarium venenatum*) is very low in fat and is useful for people who want to reduce their intake of saturated fat from animal products.

The disadvantages are as follows:

- Bacteria and fungi can be infected by viruses. If this happens then the production plant is shut down and sterilised.
- If other bacteria enter the fermenter they may compete with the bacteria being cultured, so yields are not as high. The product may need more purification treatment, so costs increase.
- It is often difficult to market a novel food and sales may be low; there may be confusion between mycoprotein and genetically modified organisms. *F. venenatum* has never been genetically modified.
- *F. venenatum* is high in nucleic acids (DNA and RNA), which have to be removed before the food is safe. Purines in the nucleic acids are metabolised in the body to uric acid, which can cause gout.
- Processing costs can be high and reduce the profit margin for foods made from microorganisms.

Batch culture and continuous culture

In a **batch culture**, a starter culture of yeast or bacteria is added to an appropriate substrate and the fermentation runs for the length of time needed for the organism to complete the sigmoid growth pattern. At the end of that time the fermenter is emptied, sterilised and the whole process started again with a new batch.

Continuous culture is used in the manufacture of Quorn™. It involves maintaining a culture of the filamentous fungus *Fusarium venenatum* in the exponential phase by supplying nutrients continuously. Fermentation broth containing fungal hyphae (the product) is removed continuously.

Batch and continuous culture are compared in Table 12.

Sensors inside fermenters detect changes in conditions to make sure that they are kept constant or changed appropriately during the fermentation to give the best growing conditions for the microorganism in order to maximise the yield. The respiration of the microorganisms generates heat, which can increase the temperature so that enzymes are denatured. Most fermenters have cooling jackets to remove excess heat. Enzymes are sensitive to changes in pH so this is adjusted by adding acids or alkalis. Sterile air is pumped in to mix the contents and provide oxygen for respiration for organisms that respire aerobically.

Products may undergo filtration, purification and packaging, collectively called **downstream processing**.

Table 12 Comparison between batch culture and continuous culture

Feature	Batch culture	Continuous culture
Fermenter	Closed	Open
Nutrients	Added at the start of fermentation	Added continuously
Product	Collected at the end of fermentation	Collected continuously throughout the fermentation
Growth phase of microorganism	Exponential phase is short	Organisms kept in the exponential phase
Examples of products	Wine, beer, yoghurt, cheese, enzymes for washing powders and penicillin (in a fed-batch culture)	Production of Quorn™ (mycoprotein); production of human insulin by GM yeast
Advantages	Easy to control conditions; produces secondary metabolites	Greater productivity; no need to empty and sterilise the fermenter
Disadvantages	Large vessels needed; waste builds up	Difficult to control conditions; no secondary metabolites produced

Knowledge check 28

Explain why the making of yoghurt, beer, bread and cheese are biotechnology industries.

Bioremediation

Bioremediation is the use of organisms to remove toxic materials from the environment. These might be heavy metals, such as mercury, lead, cadmium, zinc, chromium and nickel. They can accumulate in the environment and increase in concentration in food chains.

Hydrocarbons are common in petroleum and are relatively simple for microorganisms to break down. The addition of nutrients to areas affected by oil spills encourages growth of bacteria that metabolise the hydrocarbons in crude oil. Sources of nitrogen, phosphorus and trace elements that microorganisms require as cofactors can be added. The use of bacterial fertilisers to boost growth appears to have few adverse ecological effects, because the end products of bacterial metabolism of hydrocarbons are carbon dioxide, water and biomass.

The bacterium *Desulfovibrio desulfuricans* converts sulfate into hydrogen sulfide (H_2S). This bacterium uses acetate, lactate, pyruvate and more complex organic compounds as energy and carbon sources.

Immobilised enzymes

In some biotechnology processes enzymes are used rather than whole organisms. This is when a one-step process is required, such as the production of fructose from glucose by the enzyme glucose isomerase or the breakdown of lactose in milk by lactase to give lactose-free milk.

These two enzymes and many more are used in large-scale commercial processes. They are immobilised by being bonded to an insoluble matrix, held inside a gel lattice such as silica gel, enclosed in tiny capsules of alginate or a partially permeable membrane. Immobilised enzymes are more stable than non-immobilised enzymes when exposed to changing conditions of pH or temperature. They are removed easily

Bioremediation The use of organisms to remove toxic materials from the environment.

Exam tip

You are likely to use immobilised lactase and glucose isomerase in your practical work. It is worth finding out more details of the reactions catalysed by these two enzymes. The specification lists four others that could be used as contexts for questions.

at the end of fermentation, so do not contaminate the product. This reduces the cost of downstream processing. They can also be reused many times.

Synoptic links

Microorganisms are cultured in bioreactors to produce useful products. The organisms may respire aerobically or anaerobically. Questions on biotechnology might require an understanding of respiration from Module 5.

Biotechnology processes provide many foods and food ingredients, for example monosodium glutamate. Questions on biotechnology and immobilised enzymes might also test knowledge about biochemical molecules and enzymes from Module 2.

Summary

- Cloning occurs naturally in plants that reproduce vegetatively. These natural clones are produced in horticulture by vegetative propagation, for example by taking cuttings.
- Clones of plants are produced in tissue culture by taking explants from leaves, stems or roots. Micropropagation involves using meristems. These clones are genetically identical and so give uniform crops. All individuals of a clone may be susceptible to the same strain of a plant pathogen.
- Natural cloning in animals occurs by embryo splitting to form identical twins. Animal cloning is done by artificially splitting embryos. Somatic cell nuclear transfer (SCNT) involves transplanting nuclei from an individual with required features to enucleated eggs, which are placed inside surrogate females. The animals produced show the desired features, but there may be problems with shortened lifespan, deformities and other health problems.
- Biotechnology is the use of living organisms (especially microorganisms), or cells taken from living organisms, to make products, such as foods and drugs.
- Microorganisms have simple requirements, have short life cycles, can be kept in large containers (fermenters), do not need large spaces for cultivation and can be genetically modified more easily than animals or plants. Aseptic technique is used to avoid contamination by competitors and parasites.
- The growth of a microorganism in a closed (batch) culture is sigmoidal, with lag, exponential, stationary and decline phases. Organisms in continuous culture are held in the exponential phase. Limiting factors, such as space, nutrients, temperature and pH, influence the growth and are maintained at optimum values in fermenters to maximise yields.
- Primary metabolites (e.g. amino acids, ethanol) are produced during growth of an organism; secondary metabolites (e.g. penicillin) are not required to support growth.
- Conditions in fermenters, such as nutrient concentration, temperature and pH, are manipulated to maximise yields.
- Enzymes are immobilised by encapsulating in alginate beads or attaching them to surfaces. This allows enzymes to be reused in large-scale production processes, prevents them from contaminating products and makes them more heat stable.

Ecosystems

Key concepts you must understand

An **ecosystem** is a place that consists of a community of organisms, the abiotic (physical) factors that influence the community and the interactions between the organisms. Ecosystems range in size from an open ocean or the huge expanse of the Great Barrier Reef to ponds, dead trees, playing fields and small pools of seawater on rocky shores at low tide.

Ecosystems are dynamic, with energy flowing from the Sun through autotrophs to heterotrophs and decomposers. Elements, such as carbon, nitrogen, sulfur and phosphorus, are continually cycled within ecosystems between the living organisms and the abiotic environment. Energy flow and nutrient cycling are influenced by human activities in artificial ecosystems, such as arable and livestock farms.

Exam tip

You should revise the ecology that you studied in Module 4 while learning the topics in this section. Expect to be asked about fieldwork techniques in the exam papers.

Key facts you must know

A playing field is an ecosystem that you may study. Table 13 lists some of the biotic and abiotic factors that influence playing fields.

Table 13 Some biotic and abiotic factors that influence a playing field ecosystem

Biotic factors	Abiotic factors
Grazing by rabbits, insects and other herbivores; mowing	Soil type, for example loam, clay, sand
Trampling by people	Edaphic (soil) factors: pH, temperature, water content, air content, mineral nutrient content, humus content
Competition between grasses and weeds, such as dandelions and plantains	Rainfall and drainage
Application of herbicide sprays	Air temperature, humidity
Application of fertilisers	Light intensity, light duration

Transfer of biomass and energy

Food chains and **food webs** are ways in which the flows of biomass and energy in an ecosystem are depicted. Food chains tend to be short. Much of the energy that enters the organisms at one trophic level is used by those organisms and is, therefore, not available to the next trophic level. Energy is 'lost' during respiration and in heat loss. Only a small percentage of the energy that enters a trophic level becomes stored in the bodies of the organisms in that trophic level to be eaten by those of the next.

Energy is not recycled — it leaves the ecosystem as infrared radiation that warms the atmosphere. A moment's thought should tell you that it cannot be recycled as sunlight. The flow of energy and biomass between trophic levels is therefore not very efficient, as outlined below.

Energy flow from plants to primary consumers

Little of the light energy that strikes plants is used in photosynthesis. Reasons for this include the following:

- Light is reflected from the surfaces of leaves.
- Light passes straight through leaves.
- It is too cold for the chloroplast enzymes to function efficiently.
- Carbon dioxide is in short supply.

The **gross primary productivity** is the energy trapped by plants in photosynthesis (per unit time). Some of this is used in respiration. The remainder, which is available for maintenance and new growth in the plants, is the **net primary productivity**. At best, crop plants make available to us 5% of the energy that strikes their leaves. In natural ecosystems, the percentage of energy available to primary consumers is even lower. Not all of the energy in plants reaches the primary consumers as some plant matter is not eaten, some cannot be digested and much will die and decay rather than be eaten by grazers. The respiration in all organisms is only about 30% efficient at transferring energy to ATP; the rest is transferred to the environment as heat.

Energy flow from primary consumers to secondary consumers

The energy input to primary consumers, such as antelopes, is equivalent to the energy content of all the grass and other plants that they eat. The energy in antelopes that is eaten by predators is the energy transferred to the next trophic level. At best, about 10% of the energy entering the primary consumer trophic level is passed to the secondary consumer trophic level. In this example, antelopes use energy:

- in keeping warm (especially at night)
- in moving about in search of food
- in producing gametes and mating
- in developing and rearing young

They also lose energy as heat during digestion of their food.

In addition, predators do not:

- consume all of an antelope's body
- digest and absorb all of the antelope's flesh that they ingest

In studies of natural and artificial ecosystems ecologists take samples, find the energy content of biological materials, examine faecal contents and gut contents. This gives an idea of the quality and quantity of material produced by autotrophs and eaten by heterotrophs.

The efficiency of primary producers is calculated from the light energy striking the plants and from the energy they trap in the products of photosynthesis. It is usual to measure the photosynthetically active radiation (PAR) because plants do not absorb all the wavelengths of the visible spectrum (see the absorption spectrum on p. 40 of the guide to Module 5).

> **Exam tip**
>
> Remember that 'energy is neither created nor destroyed'. Energy cannot be 'lost' and disappear completely. If you write 'energy is lost…' you mean it is transferred so that it is of no use to organisms.

The energy content of biological materials is determined by burning them in oxygen within a piece of apparatus known as a calorimeter. The increase in temperature is recorded and from this the energy content in kilojoules is calculated.

Ecological efficiency is the efficiency of energy transfer between trophic levels. It is calculated by comparing the energy consumed by a trophic level with the energy consumed by the next trophic level. Between secondary and tertiary consumers, this is calculated as follows:

$$\frac{\text{energy consumed by tertiary consumers}}{\text{energy consumed by secondary consumers}} \times 100\%$$

Arable farmers and growers attempt to make net primary productivity as high as possible. Livestock farmers do the same for secondary productivity. Table 14 shows how human activities can alter the efficiency of biomass and energy flow.

Table 14 Some ways in which humans manipulate energy flow in artificial ecosystems

Method	Crop plants (producers)	Livestock (primary consumers)
Maximise energy input	Optimum planting distances between crop plants Provide light for greenhouse crops on overcast days	Provide good-quality feed
Maximise growth	Provide water (irrigation) and fertilisers (containing NPK and other elements, e.g. S) Selective breeding for fast growth	Provide food supplements (e.g. vitamins and minerals) Selective breeding for fast growth
Control disease	Fungicides	Antibiotics and vaccines
Control predation	Fencing to exclude grazers (e.g. rabbits, deer) Use pesticides to kill insect pests, nematodes, slugs, snails etc.	Extensive systems (ranching) Control predators such as wolves and foxes Intensive systems Keep animals protected from predation in sheds
Reduce competition	Plough and use herbicides to kill weeds	Control competitors such as rabbits and deer
Reduce energy loss	Breed plants that maximise energy storage in edible products, such as seeds and fruits	Keep animals in sheds — less energy is used in movement and maintaining body temperature

Decomposition and nutrient cycling

Detritivores, such as earthworms, termites and woodlice, ingest dead matter (detritus) and shred it, increasing its surface area. The material they egest in their faeces has a large surface area, providing easy access for bacteria and fungi. These decomposers secrete enzymes that catalyse the hydrolysis of large, insoluble organic molecules into small, soluble molecules that they absorb through carrier proteins using facilitated diffusion and active transport; the substances are then used in

Exam tip

This section on energy and biomass transfer is an opportunity for questions that test your maths skills.

Knowledge check 29

The energy consumed by secondary consumers in an ecosystem is $3610\,\text{kJ}\,\text{m}^{-2}\,\text{y}^{-1}$. The energy consumed by tertiary consumers is $135\,\text{kJ}\,\text{m}^{-2}\,\text{y}^{-1}$. Calculate the efficiency of energy transfer between the secondary consumers and the tertiary consumers in this ecosystem.

metabolism. Carbon in these compounds passes to the environment as carbon dioxide when the compounds are respired. This 'unlocks' carbon from dead organic matter, making it available for carbon fixation in photosynthesis (Figure 25).

The carbon cycle

Life is dependent on carbon as it is present in all biochemical molecules. The main processes involved in cycling carbon are:

- carbon fixation in photosynthesis
- transfer of biomass along grazing and detritus food chains
- respiration of decomposers to release carbon dioxide
- decarboxylation of organic compounds during the link reaction and Krebs cycle in respiration of organisms
- compression of undecayed organic remains to form fossil fuels
- combustion of wood and fossil fuels (peat, oil, coal and gas)
- conversion of carbon dioxide into carbonates in shells of marine animals
- deposition of calcium carbonate in animal shells and skeletons on the ocean floor and the formation of limestone and chalk

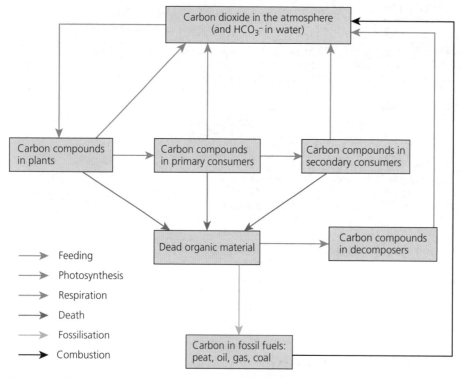

Figure 25 The carbon cycle. Use the key to identify the exchanges from the list above

Carbon in organic materials, soils and rocks forms **carbon sinks**, which 'lock up' carbon for long periods of time. Carbon sinks reduce the concentration of carbon dioxide that might otherwise be in the atmosphere contributing to a greater greenhouse effect and much higher temperatures. Examples of carbon sinks are long-lived trees, peat bogs, fossil fuels, limestone and chalk.

The nitrogen cycle

Nitrogen is an important element because it is part of many biological molecules. Most organisms take in nitrogen that is already 'fixed' in that it is combined with another element, such as hydrogen or oxygen. Plants absorb ammonium ions or nitrate ions, which are simple forms of fixed nitrogen. Animals obtain nitrogen by eating food containing proteins and other complex nitrogenous compounds. There is only a limited supply of fixed nitrogen in natural ecosystems and the supply of nitrate ions for plants depends on the action of microorganisms that use nitrogen compounds. Part of the nitrogen cycle is shown in Figure 26.

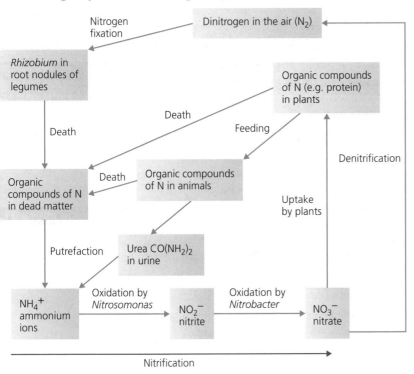

Figure 26 Part of the nitrogen cycle. Note the importance of bacteria in recycling nitrogen compounds

Use Figure 26 to find the features of the nitrogen cycle that follow.

Features involving plants:

- Absorb nitrate ions (NO_3^-) from the soil.
- Convert nitrate to ammonium ions (NH_4^+) and use them to make amino acids by the process of amination (remember that amino acids have the amino group $-NH_2$).
- Assemble amino acids into polypeptides and proteins.
- Lose leaves and die, forming dead matter (e.g. leaf litter) that contains proteins.

Features involving herbivores:

- Eat plants and digest proteins to amino acids.
- Use amino acids to make proteins (e.g. haemoglobin, collagen).

- Break down excess proteins into amino acids and deaminate them to form ammonium ions.
- Convert ammonium ions to urea in the ornithine cycle.
- Excrete ammonium ions and urea in urine.
- Produce dung that contains proteins.
- Die, leaving bodies that contain proteins.

Features involving putrefying bacteria:
- Digest proteins in dead organisms into amino acids.
- Deaminate amino acids to form ammonium ions.

Features involving nitrification by nitrifying bacteria:
- *Nitrosomonas* oxidises ammonium ions to nitrite ions to provide itself with energy.
- *Nitrobacter* oxidises nitrite ions to nitrate ions to provide itself with energy.
- Nitrate ions released by *Nitrobacter* are available for plants to absorb and so the cycle is completed.

Features involving denitrification:
- Nitrate ions are converted into dinitrogen (N_2) by denitrifying bacteria.
- Denitrification occurs in waterlogged, anaerobic soils, decreasing their fertility.

Nitrogen fixation

There is a huge quantity of nitrogen in the atmosphere. About 80% of the air around us is nitrogen, but it is in the form of the gas, dinitrogen (N_2), in which there is a strong triple bond between the nitrogen atoms, ($N{\equiv}N$). Some bacteria, such as *Rhizobium* and *Azotobacter*, fix nitrogen by using energy (from ATP) and the enzyme nitrogenase to break the triple bond and combine nitrogen atoms with hydrogen to form ammonium ions (NH_4^+).

Rhizobium lives in a mutualistic relationship with legumes, such as peas and beans, which provide it with energy in the form of sugars. Some genera of prokaryotes fix nitrogen but do not live within the body of another organism. *Azotobacter* is an example. *Rhizobium* and *Azotobacter* are examples of nitrogen-fixing bacteria. Question 1 on p. 86 is about *Azotobacter*.

> **Exam tip**
>
> Make a large flow chart of the nitrogen cycle and use it to show links to other topics. Identify all the biological molecules that contain nitrogen and include their roles in organisms.

Nitrification The conversion of ammonia to nitrate ions by bacteria.

Denitrification The conversion of nitrate ions to nitrogen gas (N_2).

Nitrogen fixation The conversion of nitrogen gas (N_2) into ammonia.

Knowledge check 30

Explain the importance of detritivores and decomposers in ecosystems.

Knowledge check 31

Make a table to show the roles of microorganisms in the nitrogen cycle.

Succession

When land becomes available for colonisation for the first time a process of **primary succession** begins in which the community changes over time until eventually reaching the **climax community** typical of the local area. The pioneer community that develops first depends on the nature of the environment — for example, the type that occurs on sand that is blown into dunes is different from that on the rock left behind by retreating glaciers.

Plants and animals in the different communities show adaptations to survival. Those in pioneer communities show adaptations to low nutrient and water availability and harsh abiotic factors. Pioneer plants often self-pollinate, which is an advantage because individual plants are often widely scattered.

Communities modify the abiotic environment with each successive community, making the environment more favourable for colonisation by other species. Pioneer species are outcompeted by grasses and low-growing, nitrogen-fixing plants. Over time a deeper soil develops, which allows the establishment of shrubs and trees, and grasses are often outcompeted.

Most climax communities are dominated by large trees. The mix of species is determined by abiotic factors, such as climate, altitude and soil type. Productivity is low because trees have much woody biomass that does not carry out photosynthesis. Earlier stages in the succession are much more productive as they are dominated by herbaceous plants with little woody tissue.

A **deflected succession** occurs when the changes do not reach the climax community. This is often because some factor, biotic and/or abiotic, favours an earlier stage in the succession. Intense grazing maintains a grassland ecosystem, such as that found on chalk grassland in the UK. Grazing animals, such as sheep, eat the seedlings of any species that could grow into larger plants overshadowing and outcompeting the plants that grow in vegetation that is never more than a few centimetres in height. Controlled burning of moorland promotes the growth of heather and prevents succession turning it into woodland.

An ecosystem that shows primary succession is a good place to measure the distribution and abundance of organisms using frame quadrats in a belt transect. The results can be compared with a more uniform ecosystem, such as a playing field, where random sampling with quadrats is more appropriate.

Synoptic links

The basis of autotrophic nutrition is photosynthesis, which was covered in Module 5. Energy losses occur in respiration. Respiration is only 30–40% efficient in transferring energy from organic compounds to ATP. The rest is lost heating the body and/or the surroundings. It is only energy trapped in biological molecules eaten by the next trophic level that is passed up the food chain. Different respiratory substrates have different energy values (see p. 72 of the student guide for Module 5).

Primary succession
Progressive change in the communities that occupy an area of land that has not been colonised before, such as a new sand dune.

Climax community
The final community in a succession. The type of community is usually determined by the climate in the area concerned.

Deflected succession
An intermediate community in a succession, often with high productivity, maintained by a factor such as grazing or mowing.

Exam tip

In the second guide in this series you will find details of fieldwork sampling techniques, such as belt transects, that you need to know about for this section of Module 6.

Summary

- An ecosystem is a dynamic system consisting of a community of organisms and all the interactions between organisms and the biotic and abiotic factors of their environment.
- Biotic factors are interactions between organisms, such as competition within species (intraspecific) and between species (interspecific), and predation. Abiotic factors are physical factors that influence communities, such as soil temperature and pH.
- Ecosystems vary in size between, say, the open ocean and individual rock pools.
- Light energy is transferred into chemical energy by producers (green plants and some prokaryotes), making energy available to consumers and decomposers. Producers manufacture biomass, which becomes available to grazing and detritus food chains. Producers, consumers and decomposers are trophic levels in food chains.
- Energy and biomass transfers between trophic levels are determined by comparing the energy and biomass taken in by one trophic level with the energy and biomass available to the next.
- Farmers and growers use techniques such as applying fertilisers and using protected environments to improve the flow of energy through artificial ecosystems.
- Decomposers break down organic material into simple inorganic compounds such as carbon dioxide and ammonia.
- Producers recycle carbon by fixing it in photosynthesis, so organic compounds become available to other trophic levels.
- Ammonia released by decomposers is oxidised by nitrifying bacteria to nitrite ions and then to nitrate ions. Denitrifying bacteria convert nitrate ions to dinitrogen (N_2). Nitrogen-fixing bacteria fix dinitrogen as ammonia.
- A primary succession is the sequence of changes from bare ground to a climax community, for example from a lake to woodland.
- A deflected succession is prevented from proceeding to a climax community by activities such as grazing, mowing or burning.
- Line transects, belt transects and frame quadrats are used to study the distribution and abundance of organisms in ecosystems.

Populations and sustainability

Key concepts you should understand

A population is a group of organisms of the same species that live in a defined area at the same time. The size of a population can be determined in a variety of ways, such as direct counting of all the individuals, as may happen with a very small population — for example, all the elephants in a small nature reserve. More often populations are sampled to estimate their numbers.

Resources taken from the environment include timber and fish. Sustainable resources are those that are replaced by natural processes. Ecosystems can be managed to ensure a constant supply.

Key facts you must know

Figure 24 on page 48 shows a sigmoid growth curve for a population of bacteria in a simple ecosystem with no competitors, predators or parasites. Similar growth occurs when an organism is released into a new environment where there are few or no factors to limit its growth. Animals such as rabbits, rats, goats, cats and mice show this when introduced into new environments. The time span is longer than in Figure 24 but the pattern is the same.

Population growth slows eventually because limiting factors provide **environmental resistance** and the population has reached its **carrying capacity** — the maximum number of individuals that an environment can sustain. Many populations rise to a maximum and then fluctuate about a mean. Some populations show J-shaped growth curves with a population explosion followed by a crash when resources are exhausted. Two limiting factors for population growth are competition for resources and predation.

Intraspecific competition is competition between members of the same species. This is intense because individuals require the same resources and have the same methods for obtaining them.

Interspecific competition is competition between individuals of different species. This can happen when an organism migrates to, or is introduced into, a new environment. It is likely that there are no available niches and so there is competition between the invading species and the existing species. This has happened in the UK with the introduction of the North American signal crayfish, *Pacifastacus leniusculus*, into rivers and streams. It has competed successfully with the native crayfish, *Austropotamobius pallipes*, reducing both its distribution and abundance in England and Wales.

Predators and prey

The relationship between predators and their prey has been studied in simple ecosystems and also in the laboratory. In these examples the predator has only one prey species and the prey species has only one predator.

The cyclamen mite, *Phytonemus pallidus*, is a pest of strawberry crops in California. Infestations of these mites can be controlled by a predatory mite, *Typhlodromus reticulatus*. In a laboratory investigation, the two species of mite were released on some strawberry plants. The numbers of both types of mite were recorded over a 12-month period. The results are shown in Figure 27.

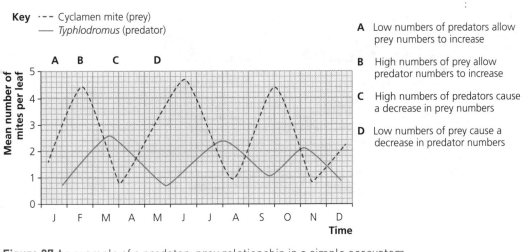

Key --- Cyclamen mite (prey)
     —— *Typhlodromus* (predator)

A Low numbers of predators allow prey numbers to increase

B High numbers of prey allow predator numbers to increase

C High numbers of predators cause a decrease in prey numbers

D Low numbers of prey cause a decrease in predator numbers

Figure 27 An example of a predator–prey relationship in a simple ecosystem

From studies like these it is commonly assumed that the population of prey species is controlled by predators, but it is far more likely that the effect of predators on prey numbers is small. In most ecosystems there are complex food webs — one predator species does not feed on only one prey species and each prey species has more than one predator. The quality and quantity of plant food available to the prey species controls their numbers, and the numbers of prey species control those of the predator. Many small mammal species show 10-year cycles determined by the availability of food plants. It is these cycles that probably have most effect on populations of their predators.

Conservation

Many natural and man-made ecosystems are at risk from human activities. People often think that natural and man-made ecosystems should be preserved for future generations. They confuse **preservation**, which involves keeping ecosystems and species exactly as they are now, with **conservation**, which involves managing the environment. Conservation involves removing vegetation to prevent succession, culling animals to prevent overpopulation, restoring habitats and creating new ones. Conservation recognises that ecosystems are dynamic and require understanding and managing, rather than simply preserving. It also involves proactive methods to maintain or increase biodiversity.

Preservation
Activities that maintain populations, communities or ecosystems in exactly the same state for the future.

Conservation
Maintaining populations and/or ecosystems by managing the environment.

Biological resources

The idea of **biological resources** can be considered at different levels. Obvious resources are those that are harvested from the wild, such as fish from our oceans and timber taken from natural forests, and those that are taken from managed ecosystems, such as fish farms and timber plantations. Ecosystems themselves are biological resources because they provide services such as decomposing wastes, recycling nutrients (Figures 25 and 26) and providing drinking water. Table 15 summarises some of the reasons for conserving these resources.

Table 15 Economic, social and ethical reasons for conserving biological resources

Reasons for conserving biological resources	Examples
Economic	Harvesting fish and managing plantations provide employment and the raw materials for many industries, such as fish and timber processing
Social	Many rural and coastal communities depend on forestry and fishing to provide people's livelihoods. Managed forests provide facilities for leisure activities, such as walking and horse riding
Ethical	We have a duty to care for our environment and maintain stocks of biological resources for future generations

Management of ecosystems to provide sustainable resources

Sustainable **timber** production involves two aspects:

- ensuring that all timber removed from an area is replaced
- maintaining the ecosystem and its biodiversity in areas where trees are grown for timber

Clear felling removes all the trees from a large area at the same time. This has adverse effects on the environment as the organisms living in that area either die or have to migrate, so reducing biodiversity. It also encourages soil erosion, especially if the plantation is on a hillside. To prevent this, **strip-felling** is carried out by cutting down strips of trees, thus avoiding the massive destruction of clear felling. **Selective felling** is employed to take out single trees.

After reaching a certain size, the trunks of some trees — for example, sweet chestnut, hazel and willow — are cut down to a stump. New shoots grow up from the stump and, after a few years, these can be harvested — a process known as **coppicing**. If different areas of the woodland are cut each year (rotational coppicing) the wood can be harvested without loss of biodiversity because the ground flora does not lose the protection of the trees. This also allows more light to reach the ground plants that would become extinct if the trees grew to full size with a dense canopy of leaves.

Many **fishing** industries have not been sustainable. Many populations of fish that are taken commercially have collapsed in the past. The herring industry used to be a big employer in ports around the North Sea. The British herring industry collapsed in the mid-twentieth century, as did the stocks of cod in Canadian waters in the North Atlantic. Fisheries are regulated in many parts of the world to try to make stocks sustainable. Some of these regulations involve:

- setting minimum mesh and maximum sizes for nets
- banning the use of some fishing methods, particularly those that catch non-target species ('bycatch') such as turtles and dolphins
- issuing quotas to each boat, so limiting the number or mass of fish landed
- setting minimum size of fish to be landed at ports
- restricting fishing effort, such as the time at sea or the time spent fishing
- restricting the size of fishing fleets and decommissioning fishing boats
- using methods to police regulations, for example fisheries protection vessels
- banning fishing in areas designated as exclusion zones
- banning fishing at certain times of the year, especially during breeding times

Sustainable resource
Any biological resource that is produced as rapidly as it is removed from the environment.

Knowledge check 32

Explain how it is possible to produce timber sustainably.

The conflict between conservation and human needs

Three case studies illustrate some of the issues involved in managing ecosystems to balance the conflict between conservation and human needs.

The Masai Mara National Reserve

The Masai Mara National Reserve occupies $1510\,\text{km}^2$ in southwest Kenya in East Africa. The reserve adjoins the Serengeti National Park in Tanzania. The Masai Mara is renowned as *the* place to see many of the large mammals that are characteristic of the African savanna ecosystem. There are over 90 species of mammal in the reserve and over 500 species of bird. It is perhaps most famous for the annual migration of herbivores, such as wildebeest, zebra, Thomson's gazelle and eland, into the reserve from the Loita Plains to the north and more famously from the Serengeti to the south. These herbivores move through the reserve and surrounding areas to feed on lush vegetation, including red oat grass, *Themeda triandra*, between July and October.

The human pressure on the reserve is immense. The human population of Kenya is growing at a rate of over 2% a year. The reserve is surrounded by many people, who are either unemployed or underemployed and subsist on very small sums of money each day. They encroach on the land to grow crops, make charcoal and to poach animals. Shanty towns are developing fast. There are too many cows for not enough land, and wheat fields are advancing. Human waste is being buried or dumped rather than treated properly.

These pressures have resulted in overgrazing by cattle, with a reduction in grazing available to the large herbivores. In response, areas of land to the north and east of the reserve were designated as privately owned reserves, known as conservancies. These doubled the area protected for wildlife. The land is owned by local people but leased to the conservancies, which have restricted the amount of grazing by livestock. The numbers of tourists are limited. Within a few years of good management the grazing has recovered and there is free movement of wildlife within the reserve.

The Masai Mara attracts thousands of tourists each year, who are accommodated in an increasing number of lodges in the area. The tourist traffic through the reserve is changing an area rich in wildlife into a theme park. Tourists demand many 'services' that have changed the ecosystem that they have come to visit.

The Galápagos Islands

The Galápagos Islands are in the east of the Pacific Ocean. They are isolated from mainland Central and South America and have many **endemic species** that are found here and nowhere else.

Oceanic islands, like the Galápagos, are especially sensitive to human interference. They are small and isolated, so populations often do not have many predators or competitors and are vulnerable to competition, predation and disease when their habitats are invaded by alien species. In the past, whalers killed huge numbers of marine mammals, such as the Galápagos fur seal, *Arctocephalus galapagoensis*. People have introduced alien species either intentionally or by accident, such as goats, pigs, cats, rats, mice, dogs and many plant species, none of which was originally on the

> **Exam tip**
>
> Search online for a map of the Masai Mara and watch some videos of the great migration and the human pressures on the reserve.

> **Exam tip**
>
> Find a map of the Galápagos and look at the websites of the Galápagos National Park, the Charles Darwin Research Station and the Galápagos Conservancy for more information on endemic species, habitat restoration and controlling human activities to protect environmentally sensitive ecosystems.

islands. These introduced species compete effectively with indigenous species and are also grazers and predators. For example, elephant grass, *Pennisetum purpureum*, competes with the endemic daisy tree, *Scalesia pedunculata*.

Only four of the islands are inhabited by humans, but the population has increased exponentially as tourism has increased. There are many threats to the wildlife:

- The town of Puerto Baquerizo Moreno on San Cristóbal is growing fast.
- There is an increase in rubbish and sewage; disposal creates environmental problems, such as land and water pollution.
- Land is required for housing and agriculture.
- There is a pressure on drinking water and energy supplies.
- Unemployment is high, and many people have resorted to fishing for sea cucumbers and lobsters, so disturbing the marine environment.

Steps taken to reduce these effects include:

- the establishment of the Galápagos National Park by the Government of Ecuador in 1959
- research and conservation projects carried out by the Charles Darwin Research Station
- restrictions on visiting the uninhabited islands, such as Daphne Major
- limiting the areas on these islands that can be visited
- a marine reserve of 133 000 km^2, which protects against depredations by fishermen — this reserve is cared for by local people as well as by conservation organisations
- culling alien species, such as goats, and destroying invasive plants such as elephant grass
- inspecting visiting boats for alien species, to prevent colonisation
- captive breeding and reintroduction, for example giant tortoises

The Lake District National Park

The Lake District National Park covers just over 2000 km^2 in northwest England. It is characterised by high peaks (fells) and deep lakes. The park attracts 16 million visitors each year, most of whom drive into the area (many from the M6, which has improved access) and many who walk the fells for views of the surrounding peaks and the 19 major lakes, which include Coniston and Windermere.

Footpath erosion is a serious threat to mountain ecosystems and therefore to wildlife as well as the tourist industry. Walking is the most popular activity for visitors to the Lake District and 16 million pairs of feet are very damaging, particularly from those who walk on the footpaths on the high fells. In addition, severe flooding in 2009 washed away bridges and damaged footpaths. The concentrated pressure of people trampling the ground compacts the soil so that water does not infiltrate, so when it rains the water runs off the surface, removing soil particles. This leads to the death of plants and the loss of their roots, which bind soil particles together. The wind and rain remove even more soil particles and eventually footpaths become eroded gullies. With further erosion gullies can become very deep and difficult to walk on. This causes walkers to use the grass on either side of the paths, with the loss of even more plants and worsening erosion.

Exam tip

Use the website of the National Park for more information about the effects of human activities on the Lake District and how these are managed to minimise their impact.

By the late 1990s the pressure of walkers was so great that it looked as if a wide road had been built to the summit of Helvellyn, one of the highest peaks. This was an erosion scar, 8 m wide, that extended for about 300 m. Debris from eroded paths like this enters watercourses, leading to death of fish and invertebrates. Trout, salmon and the vendace, *Coregonus vandesius*, are fish that are threatened by this pollution of their spawning grounds. Erosion has completely destroyed the rare alpine 'mouse-ear' plant, *Cerastium alpinum*, and in Red Tarn, below the ridge of Helvellyn, the white schelly fish, *Coregonus stigmaticus*, the UK's rarest fish, may be at risk.

Management of this problem involves repairing footpaths and making them usable in the long term. Contractors repair some of the paths and teams of skilled rangers and volunteers repair paths on land owned by the National Trust within the park (Figure 28). More permanent solutions are used, such as using stones to create paths. As a result of this repair work, local plant populations have started to recover.

Figure 28 Walkers and water — two reasons why footpaths in the Lake District need constant repairs. Work on this footpath near Aira Force, a spectacular waterfall near Ullswater, involved putting in drainage and resurfacing

Summary

- The carrying capacity is the maximum number of a species that can be supported by an ecosystem. Biotic and abiotic factors limit the final size of populations.
- Predation, disease and interspecific and intraspecific competition are examples of interactions between populations in an ecosystem.
- The relationships between predators and their prey determine the sizes of their populations.
- Conservation, unlike preservation, is a dynamic process that involves managing ecosystems for the benefit of the long-term survival of species and sustainable production of natural resources, such as timber and fish stocks.
- There are economic, social and ethical reasons for conservation of biological resources.
- Ecosystems can be managed to balance the conflict between human needs and conservation, as in the Masai Mara in Kenya and the Lake District in the UK.
- Human activities, such as fishing and the introduction of alien species, have had harmful effects on animal and plant populations in the Galápagos Islands. Various measures are being taken to limit the effects of these activities.

Questions & Answers

Exam format

At A-level there are three exam papers. Questions in these three papers will be set on any of the topics from Modules 2–6 in the specification. In addition, there will be questions that will test your knowledge and understanding of practical skills from Module 1 and your ability to apply mathematical skills.

Your exams will be as follows:

Paper number	1	2	3
Paper name	Biological processes	Biological diversity	Unified biology
Length of time	2 hours 15 minutes	2 hours 15 minutes	1 hour 30 minutes
Total marks	100	100	70
Types of question	15 multiple-choice questions (1 mark each) and structured questions for 85 marks	15 multiple-choice questions (1 mark each) and structured questions for 85 marks	Structured questions
Synoptic questions	Yes	Yes	The whole paper is synoptic!

About this section

The first part of this section contains questions similar in style to those you can expect to find in Paper 2 (Biological diversity). The questions in the paper are based on topics from Modules 2–4 and 6. Most of the questions on pages 68–85 are based on topics from Module 6, but some require you to apply your knowledge of Modules 1–4.

The answers to the five multiple-choice questions are on page 93.

Questions similar to those in Paper 3 (Unified biology) are on pages 86–93. Paper 3 tests your knowledge of all the Modules (1–6), but concentrates on the skills you have developed during your practical work. Most of the questions on pages 86–93 are based on topics in Module 6 but they require more knowledge of Modules 1–4 than Paper 2.

The limited number of questions in this guide means that it is impossible to cover all the topics and all the question styles, but they should give you an indication of what you can expect in Papers 2 and 3.

As you read through the answers to the questions based on Paper 2, you will find answers from two students. Student A gains full marks for all the questions. This is so that you can see what high-grade answers look like. Student B makes a lot of mistakes — often these are ones that examiners encounter frequently. I will tell you how many marks student B gains for each question. The Paper 3-style questions only have model answers, similar to those for student A.

Comments

Each question is followed by a brief analysis of what to watch out for when answering the question (icon **e**). Some student responses are then followed by comments. These are preceded by the icon **e** and indicate where credit is due. In the weaker answers, they also point out areas for improvement, specific problems and common errors, such as lack of clarity, weak or non-existent development, irrelevance, misinterpretation of the question and mistaken meanings of terms.

■ Paper 2-style questions: Biological diversity

Section A Multiple-choice questions

Question 1

Which is an example of somatic cell gene therapy?

A cloning a gene in *Escherichia coli* cultured in a fermenter

B inserting a gene into an egg cell immediately after fertilisation

C inserting a gene into stem cells that are transplanted into bone marrow

D producing a human polypeptide in genetically modified hamster ovary cells   (1 mark)

Question 2

What is the role of cyclic AMP in gene expression?

A activating enzymes by binding to allosteric sites

B activating RNA polymerase by binding directly to DNA

C cutting introns from primary mRNA to form mature RNA

D folding polypeptides into specific shapes by binding to R groups   (1 mark)

Question 3

Haemochromatosis is the most common genetic disorder in northern Europe. Mutations of the *HFE* gene on chromosome 6 are the cause of this disorder. People with the disorder are homozygous recessive for the mutation. The prevalence of the disorder in the UK population is 0.4%. What is the frequency of carriers of the disorder in the UK population?

A 5.9%

B 11.80%

C 39.84%

D 80.68%   (1 mark)

Question 4

Most cultivated banana plants belong to the variety Cavendish. These plants are triploid (3*n*). What is the reason why these cannot be used in a breeding programme for resistance to black sikatoga?

A   Cavendish plants cannot be propagated.

B   Cavendish plants are sterile.

C   Cavendish plants are susceptible to black sikatoga.

D   Cavendish plants are vulnerable to very cold conditions.   (1 mark)

Question 5

Which factor limits the number of trophic levels in a food chain?

A   biomass of the autotrophs

B   efficiency of energy conversion between trophic levels

C   net secondary productivity

D   species diversity in the ecosystem   (1 mark)

Section B Structured questions

Question 6

The enzyme α-galactosidase catalyses the hydrolysis of the disaccharide lactose into glucose and galactose. This enzyme is synthesised by the bacterium *Escherichia coli* only when lactose is present. The gene for α-galactosidase is part of the *lac* operon.

(a) Explain what is meant by the term *operon*.   (2 marks)

ⓔ Here you need to remember a suitable definition. This is why it is a good idea to write your own glossary to help your revision.

(b) Explain the advantage of synthesising α-galactosidase only when lactose is present.   (2 marks)

ⓔ For this question, think of the advantages to the bacteria. Do not just think about the operon.

Using a sterile pipette, a sample was taken from a laboratory culture of *E. coli*. The sample was transferred to a large volume of sterile culture medium containing $0.001 \, \text{mol} \, \text{dm}^{-3}$ of glucose and $0.001 \, \text{mol} \, \text{dm}^{-3}$ of lactose.

(c) Explain why it is important that sterile conditions were used in this investigation.   (2 marks)

ⓔ In microbiological work it is important to use aseptic technique. You should know both the precautions to take and the reasons for them.

The population of live bacteria in the culture was determined at 15-minute intervals. The results are shown in Figure 1.

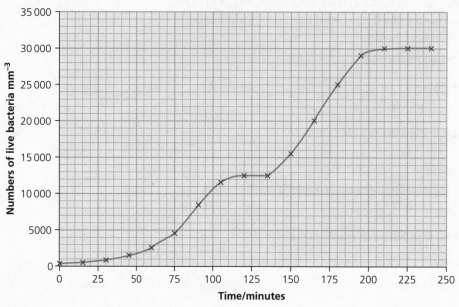

Figure 1

(d) Use the information in Figure 1 to answer the following questions.

(i) Explain what happened to the population of *E. coli* during the first 100 minutes.

(4 marks)

ⓔ The command word is *explain* not *describe*. Before you start writing your answer look carefully at the graph for the first 100 minutes and ask why the numbers increased. Think of appropriate terms to use and work out an explanation.

(ii) Explain why the population remained stable after 120 minutes and then increased after 135 minutes.

(4 marks)

ⓔ Look at the information about the *lac* operon and use your knowledge of its control.

(iii) Suggest why the number of live bacteria was counted rather than the total number of bacteria in each sample.

(1 mark)

ⓔ This is a *suggest* question, so do not think you know the answer. You have to work it out. The clue is 'live' in the question.

Student A

(a) An operon is an arrangement of genes with common regulator sections of DNA (promoter and operator regions). The structural genes in the *lac* operon are close together and are transcribed together. The regulator sections determine whether the genes are transcribed or not. A repressor substance determines whether the genes are switched 'on' or 'off'.

ⓔ **2/2 marks awarded** Student A gives a thorough answer, generalising from the *lac* operon to other operons.

Student B

(a) An operon is a regulator gene and some genes that code for enzymes.

ℯ **0/2 marks awarded** Student B has not developed the answer to explain that operons determine whether or not genes are transcribed by RNA polymerase. There is no reference to operator and promoter regions.

Student A

(b) It is a waste of energy in the form of ATP to synthesise enzymes unless the correct conditions (e.g. food) are available. It is wasteful to make enzymes when they are not wanted. Some enzymes are only needed at certain times during the life of a cell.

ℯ **2/2 marks awarded** Student A gives the reason why it is an advantage and also refers to synoptic material by mentioning ATP from Module 5.

Student B

(b) Lactose is not available, so there is no need for that enzyme. The repressor substance binds to the operator region so that the gene for the enzyme cannot be made.

ℯ **0/2 marks awarded** Student B has not developed the idea that there is no need for the enzyme in terms of not wasting energy or materials, such as the amino acids that are needed to make α-galactosidase. Student B has written 'so that the gene for the enzyme cannot be made' instead of 'so that the gene for the enzyme cannot be transcribed' or 'so that the mRNA for the enzyme cannot be made'.

Student A

(c) Aseptic technique must be used so that bacteria and spores from the air and any on the inside of the flask or in the culture medium do not grow and use up the nutrients. The numbers counted would include other species and be hard to interpret.

ℯ **2/2 marks awarded** Student A has explained the need for aseptic technique and has thought about the effect of competition on the suitability of the investigation.

Student B

(c) If the conditions were not sterile then other bacteria or fungi would contaminate the culture medium and compete with *E. coli* for the nutrients and oxygen.

ℯ **2/2 marks awarded** Student B has also explained the need for aseptic technique in terms of competition for nutrients.

Questions & Answers

> **Student A**
>
> **(d) (i)** When the bacteria are first put into the flask they are in a lag stage. During the first 45 minutes the bacteria are absorbing nutrients, making enzymes and transport proteins for membranes. Next is the log stage — this is between 45 and 100 minutes, when the bacteria multiply rapidly and the population reaches 10000 per mm^3. Then the increase starts to slow as factors become limiting and environmental resistance sets in.

ⓔ 4/4 marks awarded 'Using the data...' and 'With reference to Figure...' mean that you can quote some of the data from the graph, table or other material that is provided. In this case it is necessary to identify the growth phases, and student A has done this effectively.

> **Student B**
>
> **(d) (i)** The population shows sigmoid growth during this time. To begin with the bacteria do not divide at all as they are not used to the environment. Then there is a dramatic increase in numbers because there are no limiting factors.

ⓔ 1/4 marks awarded Student B has not quoted the data. It is possible to gain full marks for 'describe' questions without doing so, but it is more difficult. Here student B refers to sigmoid growth and gains a mark for using that term correctly. 'To begin with' and 'dramatic increase' are not precise enough to gain marks for description; 'no limiting factors' should be illustrated by at least one example.

> **Student A**
>
> **(d) (ii)** The population is stable because the glucose is now exhausted and the rate of production of bacteria by binary fission is equal to the death rate. There is a food supply available (lactose), but the bacteria do not have the lactase enzyme. Some lactose enters the bacteria, binds to the repressor substance removing the inhibition, so the gene for lactase is transcribed and translated. This didn't happen earlier because the *lac* operon is turned off when glucose is present, because it is easier to respire glucose as there is no need for an enzyme to hydrolyse glycosidic bonds.

ⓔ 4/4 marks awarded Prompted by the name of the enzyme, student A refers to the glycosidic bond in disaccharides. This is good use of material from Module 2.

> **Student B**
>
> **(d) (ii)** This stage is called the 'stationary phase'. Here the rate of growth is constant because the nutrients have run out. The number of bacteria made is equal to the number that dies. Excretory products will also inhibit the metabolism of the bacteria.

🅔 **2/4 marks awarded** Student B does not respond to the second part of the question, which asks for an explanation of the second part of the graph. Look out for two-part questions like this. Examiners may ask for a description *and* an explanation, or advantages *and* disadvantages, and embolden the 'and', so you don't miss the instruction as student B has done here.

> **Student A**
>
> **(d) (iii)** Some of the bacteria would be dead but not decomposed, so you would see them if you put a sample under the microscope.

🅔 **1/1 mark awarded** This is enough for the mark.

> **Student B**
>
> **(d) (iii)** Many would be dead, so you would not see the plateau when you plot the total numbers as bacteria are still dividing during the stationary phase.

🅔 **1/1 mark awarded** Student B has the right idea here.

Question 7

The synthesis of the blue pigment, malvidin, in flowers of *Primula* is controlled by two unlinked genes, **K/k** and **D/d**. The dominant allele, **K**, codes for the production of malvidin. No pigment is produced by plants with the genotype **kk** and their flowers are white. The production of the pigment is suppressed by the dominant allele **D**, but not by the recessive allele, **d**.

(a) State the genotypes of *Primula* plants that have blue flowers. (1 mark)

🅔 Note that there are two gene loci in the information above; both must be included in your answer. See p. 23 for advice on writing genotypes in dihybrid crosses. Always read the whole question before answering part (a). You can see how other genotypes are written in (b).

(b) Two *Primula* plants with the genotypes **KkDd** and **kkdd** were crossed. Draw a genetic diagram to show the genotypes and phenotypes of the offspring. State the ratio of phenotypes in the offspring. (5 marks)

🅔 You might be given a grid with subheadings (*parental phenotypes, parental genotypes,* etc.) to complete. If not, follow the pattern in the genetic diagrams used in this guide (pp. 18–24).

(c) (i) State the name given to the gene interaction shown in the control of flower colour in *Primula*. (1 mark)

 (ii) Suggest how allele **D** interacts with the locus **K/k** to give the results you have shown in part (b). (2 marks)

🅔 You should have identified the type of interaction while reading the introduction to this question. You should write this down on the exam paper as you read. It will help you to focus on the appropriate sections from the specification.

(d) Some students investigated the inheritance of two genes in the fruit fly, *Drosophila melanogaster*. Pure-bred flies with straight wings and dark red eyes were crossed with pure-bred flies with curved wings and brown eyes. All the F_1 generation had straight wings and dark red eyes.

In a test cross, the females of the F_1 generation were crossed with males with curved wings and brown eyes. The students expected a ratio of 1:1:1:1 in the test cross offspring.

The results were as follows:

- straight wings and dark red eyes   79
- curved wings and dark red eyes   25
- straight wings and brown eyes   23
- curved wings and brown eyes   73

A chi-squared (χ^2) test was carried out to see if the number of each phenotype was in agreement with a 1:1:1:1 ratio. The value of χ^2 was 54.48.

Table 1 shows critical values of χ^2 at different levels of significance and degrees of freedom.

Table 1 Distribution of χ^2

Degrees of freedom	Probability, p				
	0.10	0.05	0.02	0.01	0.001
1	2.71	3.84	5.41	6.64	10.83
2	4.61	5.99	7.82	9.21	13.82
3	6.25	7.82	9.84	11.35	16.27
4	7.78	9.49	11.67	13.28	18.47

(i) Using Table 1, state the conclusion that would be drawn from the results of this investigation and explain how you reached this conclusion. (3 marks)

ⓔ There are some key words you should use in answer to this type of question. Make a note of these in the margin of the exam paper and include them.

(ii) Explain why the results are different from the expected results. (3 marks)

ⓔ This is where you need to apply your knowledge of inheritance to a specific example that you will not have seen before.

(e) Milk yield in cattle is influenced by both genetic and environmental factors. Explain why it is necessary to consider environmental factors when using selective breeding to improve milk yield. (3 marks)

ⓔ Milk yield is an example of continuous variation. Think about the different environmental factors that are likely to influence the milk yield of dairy cattle. These should help to frame your answer.

Student A

(a) KKdd, Kkdd

ℯ 1/1 mark awarded Student A has given the only two genotypes that result in flowers containing malvidin. To produce this pigment there must be a **K** allele and no **D**.

Student B

(a) K–dd

ℯ 0/1 mark awarded It is perfectly acceptable to use a dash (–) in genotypes. You can write this on the exam paper when working out genetics problems. However, this question asks for the *genotypes* (plural) and so student B should have realised that more than one genotype is required.

Student A

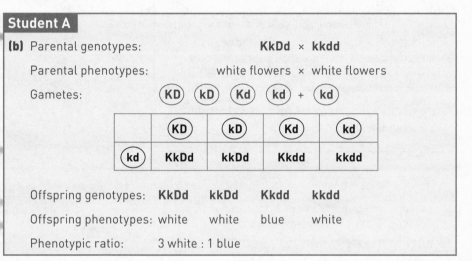

(b) Parental genotypes: KkDd × kkdd

Parental phenotypes: white flowers × white flowers

Gametes: KD   kD   Kd   kd   +   kd

	KD	kD	Kd	kd
kd	KkDd	kkDd	Kkdd	kkdd

Offspring genotypes: **KkDd**   **kkDd**   **Kkdd**   **kkdd**

Offspring phenotypes: white   white   blue   white

Phenotypic ratio: 3 white : 1 blue

ℯ 5/5 marks awarded Student A has completed all the sections of the genetic diagram (the side headings might be given on the examination paper to help you with the steps involved) and has given a Punnett square to show all the fusions that take place. This is good practice. Finally, student A has written the phenotype underneath each genotype, which is essential to gain a mark.

Student B

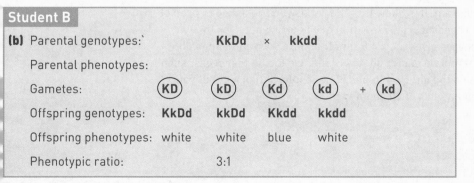

(b) Parental genotypes: KkDd   ×   kkdd

Parental phenotypes:

Gametes: KD   kD   Kd   kd   +   kd

Offspring genotypes: **KkDd**   **kkDd**   **Kkdd**   **kkdd**

Offspring phenotypes: white   white   blue   white

Phenotypic ratio: 3:1

e **3/5 marks awarded** Student B has neglected to give the parental phenotypes and has not written in the flower colours in the phenotypic ratio, so loses 2 marks. In this question the flowers without malvidin are described as white. In fact, the colour of the flowers depends on other genes. Control of flower colour in plants and coat colour in mammals are examples of features controlled by several genes that have multiple alleles (pp. 20–21). They provide many opportunities for questions on genetic control, including the topic of epistasis.

Student A

(c) **(i)** Dominant epistasis

(ii) The dominant allele could code for a transcription factor that inhibits transcription of the gene that codes for the enzyme that makes malvidin. It might act by binding directly to DNA, so blocking the action of RNA polymerase, or binding to another transcription factor, so preventing transcription. Alternatively, the inhibitor might bind to the enzyme, changing the shape of its active site so that it doesn't work.

---

e **3/3 marks awarded** Another possible explanation for part (ii) is that allele **D** codes for an enzyme that breaks down malvidin.

Student B

(c) **(i)** Epistasis

(ii) The gene could code for an enzyme that breaks down the malvidin once it is made. Or it could inhibit malvidin.

---

e **2/3 marks awarded** Student B gains the mark for part (i). In part (ii) student B should have said 'The dominant allele…' rather than 'The gene…' but in this case it is clear what is meant and 1 mark is awarded. 'It could inhibit malvidin' does not tell us much. If you say that something is an inhibitor you should always explain or suggest how it works.

Epistasis is a difficult topic — especially predicting phenotypic ratios without first drawing a genetic diagram. It is likely that the ratios you are asked to predict are those obtained when crossing organisms that are heterozygous for the two interacting genes. This always gives a 4 × 4 Punnett square with all the possible combinations of dominant and recessive alleles of the two loci. A test cross, such as that shown in (b), reveals the different types of gene interaction without you having to work out the phenotypes of 16 genotypes.

Student A

(d) (i) There is a significant difference between the observed results and the results expected by independent assortment of the alleles of the two genes (1:1:1:1). This means that the null hypothesis is rejected. This is because the value for χ^2 (54.48) is greater than the critical value of 7.82 at $df = 3$ and the 5% significance level. It is greater than 16.27, so we can be 99.9% certain that the results are not due to chance.

ⓔ **3/3 marks awarded** It is always a good idea to include the level of significance, as student A has done here.

Student B

(d) (i) degrees of freedom = 3

The critical value for χ^2 at $p = 0.05 = 7.82$. As the value of 54.48 is greater than the critical value, the null hypothesis can be rejected and, therefore, these results were not obtained simply by chance.

ⓔ **3/3 marks awarded** The 'conclusion' mentioned in the question means that you must say whether the null hypothesis is accepted or rejected.

Student A

(d) (ii) The two genes are linked on the same chromosome so independent assortment did not happen. If it had done, then the four phenotypic classes would be present in approximately equal numbers and the value for χ^2 would be less than the critical value. The recombinant classes were obtained by crossing over between non-sister chromatids.

ⓔ **3/3 marks awarded** Student A realises that the results are obtained by crossing over between the two gene loci that are linked on the same autosome.

Student B

(d) (ii) The expected results from the test cross are in the ratio 1:1:1:1, so this would be 50:50:50:50 for the different phenotypes. The results could not have been obtained by chance — for example by the random fusion of gametes. The chi-squared test shows that there must be another explanation.

ⓔ **0/3 marks awarded** Student B has simply repeated the conclusion given in part (i), so fails to score.

Student A

(e) Milk yield is a characteristic that shows continuous variation, and is therefore influenced by both genotype and the environment. It is a polygenic feature because it is controlled by many genes. The production of sugars, proteins and fats in the milk is also influenced by the quality and quantity of food given to the cows. When breeders choose females to use in breeding programmes they must make sure that they all have the same feed. If all the cows are kept under the same conditions any differences between them will be due to the alleles of the genes that control milk yield. They can then choose the best milkers to be inseminated by the best bulls.

e **3/3 marks awarded** This is an excellent answer.

Student B

(e) Selective breeding means taking certain males and females and crossing them. Milk yield is a sex-limited feature only expressed in females. The males are tested to find those that have daughters that give plenty of milk. Females with high yield are given good conditions, so the only factor controlling their milk yield is the environment. The best females (the ones that produce most milk) have the 'best' genes and should be used.

e **0/3 marks awarded** Student B has not really answered the question. There is almost a mark for writing about 'good conditions' but the examiner would expect a more precise answer — for example, mentioning the quality and quantity of food, as student A did. This answer could have explained more about the diet, but the important point is that the environment is the same for all of the cows, so the genetic component can be assessed. Student B refers to the 'best' genes. This is not a good idea, even though inverted commas are used to show that it is not exactly what is meant. All cows have the same genes. It is the alleles of those genes that differ, with some alleles contributing more to the milk yield than others.

Question 8

Georges Bank in the northwest Atlantic Ocean had a thriving herring fishery until the mid-1970s. Figure 2 shows changes in the stock of 2-year-old and 4-year-old herring, *Clupea harengus*, between 1967 and 2003.

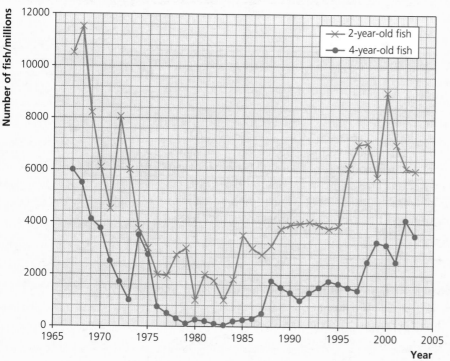

Figure 2

(a) Calculate the percentage increase in the 2-year-old stock between 1983 and 2000. Show your working.

(2 marks)

e Use a ruler to take intercepts from the graph to help you read the figures accurately.

(b) Describe the changes to the 4-year-old stock of herring between 1967 and 2003. (4 marks)

e Draw vertical lines on the graph where there are significant changes in the numbers of fish caught.

(c) There was hardly any commercial fishing on Georges Bank after the collapse of fisheries in 1977 until 1995. Suggest the long-term ecological advantages of restricting fishing in certain areas of the sea.

(5 marks)

e This question is testing your knowledge of the interactions that occur in ecosystems.

(d) Explain how stocks of wild fish are conserved so that a sustainable resource can be maintained.

(5 marks)

e Your answer should show understanding of the term *sustainable* and be more than just a list of methods to prevent overfishing.

Student A

(a) 9000 – 1000 (millions of fish) = increase of 8000 (million)

percentage increase = $\dfrac{8000}{1000} \times 100 = 800\%$

ℯ **2/2 marks awarded** Correct, for full marks.

Student B

(a) $\dfrac{1000}{9000} \times 100 = 11.1\%$

ℯ **0/2 marks awarded** Student B has made a common error in calculating the change as a percentage of the final number rather than the initial number. Sometimes people do not realise that a percentage change can be greater than 100%, cross out the correct answer and give an incorrect answer like 11.1%. Remember 'change over original times 100'. If student B had remembered this, the working would have been given and 1 mark could have been awarded even if the answer was not calculated correctly.

Student A

(b) There is a steep decrease in the number of fish caught between 1967 and 1973 from 6×10^9 to 1×10^9. In 1974 the number increased to 3.5×10^9 (an increase of 250%) and then decreased to almost zero until 1983. From 1983 onwards the numbers increased with fluctuations to about 4×10^9 in 2002–2003.

ℯ **4/4 marks awarded** Descriptions of data shown in graphs and tables should always be accompanied by figures. Questions often start with a straightforward statement of a figure or a percentage change so that you are given the cue to use the data in questions that follow. Student A has calculated another percentage change, which is good practice.

Student B

(b) The numbers fell dramatically after 1967 to 1973 and then they peaked in 1974. The population crashed and only later did it increase again.

ℯ **2/4 marks awarded** Student B has not used any data from the graph. Always use a ruler to help read and interpret graphs. In this case position the ruler against the vertical axis at 1967 and move it across the graph. Mark where key changes occur and write down the appropriate figures. Student B gains 2 marks for the overall description.

Student A

(c) The population numbers increase as species migrate into the area and recolonise the sea bottom. The species diversity increases, the food web becomes more complex as new prey species and new predators migrate into the area and the ecosystem becomes more stable. There is an increase in numbers of reproductive fish so that more are involved in spawning. As a result there is an increase in the fish stock.

ℯ **5/5 marks awarded** It is important to remember that you are expected to know about ecological principles as well as methods to make fishing sustainable. The question is quite open, so you can write about biodiversity, food webs and the interrelationships between organisms.

Student B

(c) If fishing boats are excluded from an area, then the numbers of fish should increase because none are being taken. If there are fish in the surrounding area they may migrate into the area and start to breed.

ℯ **1/5 marks awarded** Student B gains 1 mark for explaining that fish may migrate into the area but they will only do so if food is available.

Student A

(d) The methods used for each fish stock depend on population numbers. The regulator has to set the maximum number or biomass (tonnage) of fish that can be taken. Ways to regulate fishing include setting minimum mesh and maximum sizes for nets and banning the use of some fishing methods, particularly those that catch non-target species (bycatch) such as turtles and dolphin. They can also issue quotas to each boat, so limiting the number or mass of fish landed.

ℯ **5/5 marks awarded** This is a clear and concise answer.

Student B

(d) Fish stocks are affected by overfishing. When many fishing boats catch many fish there are few fish left to keep the population going, especially if they catch the bigger fish that are the best at reproduction. When a stock decreases below a certain level it may never recover, so to stop that happening governments have fishing regulations, such as not fishing at some times of the year.

ⓔ 0/5 marks awarded Student B has not answered the question. The answer deals with reasons why fish stocks collapse not ways to conserve them for the future. Answers to questions like this one must be supported with good examples. Writing these as a bulleted list is a good idea so long as each point is made clearly enough to show that it will conserve numbers of fish. Similar questions could be asked about the sustainability of timber. Remember the definition of sustainable resources from p. 63 and use it in answers to questions on biological resources (fish and timber).

Question 9

The gene for the β-polypeptide of human haemoglobin was isolated from chromosome 11. The gene was sequenced. The sequencing process began with a length of DNA isolated from chromosome 11.

(a) Outline the principles involved in sequencing this gene. (4 marks)

ⓔ There is no need to describe in detail any one method of gene sequencing. The question is asking for the *principles* involved, as described on p. 39.

Genes that code for the β-polypeptide of haemoglobin of seven other species of primate were sequenced. Figure 3 shows relationships between the eight species based on the data obtained.

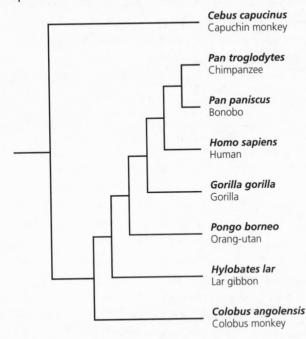

Cebus capucinus
Capuchin monkey

Pan troglodytes
Chimpanzee

Pan paniscus
Bonobo

Homo sapiens
Human

Gorilla gorilla
Gorilla

Pongo borneo
Orang-utan

Hylobates lar
Lar gibbon

Colobus angolensis
Colobus monkey

Figure 3

(b) With reference to Figure 3, explain how data from gene sequencing can be used to assess the evolutionary relationships between different species. (4 marks)

ⓔ 'With reference to...' means that you can give information from the diagram in your answer. These are usually easy marks, so spend some time looking carefully at the information provided.

(c) *Cepaea nemoralis* is a species of snail found widely in the UK. The presence or absence of bands on the shell is controlled by a single gene, **B/b**. The allele for unbanded, **B**, is dominant to the allele for banded, **b**. In a population of these snails, 36% showed the banded phenotype. In parts of the UK, the proportion of snails with bands in populations of *C. nemoralis* has remained constant for many years.

The Hardy–Weinberg principle is used to estimate the frequency of alleles and genotypes in populations. The Hardy–Weinberg equations are:

$p + q = 1$ and $p^2 + 2pq + q^2 = 1$

Use these equations to calculate the frequency of the allele for unbanded in this population. Explain how you arrive at your answer.

(3 marks)

ⓔ Remember that examiners usually give marks for showing the working in calculations. In some calculations you may be given full marks even if you do not include any working. Here you must show your working. Write out an explanation *in words*. Do not just write down the numbers.

Student A

(a) Gene sequencing involves determining the base sequence in a complete gene like that for the β-polypeptide of haemoglobin. The process is a modification of semi-conservative replication in which DNA polymerase uses lengths of DNA from the gene as templates for assembling nucleotides. The four different nucleotides are 'tagged' in some way so that they are identified as they are added to the newly synthesised strand of DNA.

ⓔ **4/4 marks awarded** Student A has answered the question as set by outlining the principles involved in gene sequencing. These principles are essentially the principles of replication that you learned about in Module 2.

Student B

(a) When a gene is sequenced it is translated from a base sequence into an amino acid sequence. Each amino acid is coded by a codon or codons that consist of three bases, for example AAA, TTT etc. To do this for a gene you find the start codon (AUG) and then look at the genetic code to find the amino acids that correspond to each of the codons. The sequence is the primary structure of the polypeptide coded by the gene you have just sequenced.

ⓔ **0/4 marks awarded** Student B has misinterpreted the question and thought that 'sequencing' meant translating a sequence of nucleotides or bases into a sequence of amino acids using the genetic code shown in Table 1 on p. 7. The answer is a bit misleading as one of the examples of a codon is TTT, which is a DNA triplet. The start codon (AUG) is in mRNA. If you are writing about nucleotide (base) sequences always make it clear whether you are writing about DNA or mRNA. Also be careful about the two strands of DNA — the coding and non-coding (template) strand. Even though the answer includes correct information, student B has not answered the question and so gains no marks.

Student A

(b) The branching points in the diagram indicate where there are significant differences between the gene sequences. The polypeptides of haemoglobin function to transport oxygen, but there are differences in the base sequences that code for the order of the amino acids and also in the non-coding regions (e.g. introns — non-coding regions within a gene). Substitution mutations occur so that the sequences are different. These may either code for different amino acids or be silent mutations (often the third base is the one that changes and the new triplet codes for the same amino acid as the original). These will be neutral mutations as all organisms have functional haemoglobin.

So where the branching point is to the right the species are closely related as they have most of the base sequence in common (e.g. chimpanzee and bonobo). Where the branching point is to the left, the species are less closely related as there are more differences in the gene sequence (e.g. *Gorilla* and *Cebus*). This information is compared with information about other genes, non-coding regions of DNA and protein sequences to work out the evolutionary history (phylogeny) of the primates.

@ **4/4 marks awarded** This phylogenetic diagram has been drawn using limited data from one gene. However, looking at the base sequences of individual genes usually confirms the relationships between species that have been discovered from other sources of information such as morphology, anatomy, physiology and protein sequencing. Student A has written far too much for a 4-mark question and in doing so risks running out of time.

Student B

(b) This shows that humans and chimps are more closely related to each other than either is related to the other primates, for example gorilla. The monkeys are not closely related at all. The differences between gene sequences show that mutation must have occurred but the protein that the gene codes for still works as an oxygen carrier.

@ **0/4 marks awarded** Student B describes what the diagram shows but does not explain how gene sequences are used and so fails to score.

Student A

(c) Banded is the recessive phenotype, so 36% are homozygous recessive.

36% = 0.36

$\sqrt{36}$ = 0.6 = q (frequency of the recessive allele)

p = frequency of the dominant allele

p = 1 – 0.6 = 0.4

🅔 **3/3 marks awarded** This is correct, with a suitable explanation. Note that this question is not like the usual calculation questions that have 2 marks. Here, the examiner asks for an explanation, rather than giving the usual instruction to 'show your working'. It might help the answer to use symbols for the alleles, although it is not necessary.

Student B

(c) If 36% are banded, then 64% are unbanded.

Unbanded is dominant, so unbanded snails are either homozygous dominant or heterozygous.

$p^2 + 2pq + q^2 = 1$

We know that $q^2 = 0.36$, so q must be 0.6.

Therefore, as $p + q = 1$, p = 1 − q, so q must be 0.4.

The frequency of the dominant allele is 0.4 or 40%.

🅔 **3/3 marks awarded** Student B's method is rather long-winded, but work through it if you are unsure about how to answer questions on the Hardy–Weinberg principle. The question says that these frequencies have existed for a long time — you can see that the population is in equilibrium because you can calculate the frequencies in the next generation and find they are the same as in the existing one.

🅔 Overall, student B gains 20 marks out of 60. This sort of performance may not be good enough for an E grade. Marks were lost for a number of reasons:

- Not developing answers fully (e.g. Q.6a, Q6b, Q6di, Q.8c).
- Interpreting terms from the specification incorrectly (e.g. Q.9a).
- Not explaining terms (Q.6a).
- Describing, rather than explaining, information (e.g. Q.7e).
- Not understanding what is required in an answer (e.g. Q.6a, Q.7dii, Q.9a).
- Not following instructions carefully (e.g. Q.7a, Q.6dii).
- Not using the data provided (e.g. Q.8b).
- Not answering precisely (e.g. Q.6di).
- Missing parts of questions (e.g. Q.6dii).
- Including irrelevant material (e.g. Q.9b).
- Not learning the facts (e.g. Q.8d).
- Repeating a previous answer (e.g. Q.7dii).
- Not giving all the steps (e.g. the genetic diagram in Q.7b).
- Not using synoptic material to the best effect (e.g. Q.6b).
- Not completing a calculation correctly (e.g. Q.8a).

■ Paper 3-style questions: Unified biology

Question 1

Eucalyptus camaldulensis is an important tree crop in India. Scientists in Coimbatore, Tamil Nadu, investigated the use of *Azotobacter chroococcum* as an alternative to the synthetic auxin, IBA, for encouraging cuttings of *E. camaldulensis* to take root and become established. They made a culture of the bacterium that contained 5×10^7 bacteria per cm^3.

They set up the following treatments:

■ cuttings that were treated with IBA

■ cuttings that were treated with $5\,cm^3$ of the bacterial culture

■ cuttings that were treated with $10\,cm^3$ of the bacterial culture

■ cuttings that were not treated in any way

The cuttings were grown in vermiculite (an inert medium that only provides support) and screened after 30 days for several features. The experiment was replicated three times and the means of the results are shown in Table 1.

Table 1

Treatment of cuttings	Root length/ mm	Biomass/ mg plant^{-1}	Number of roots per plant	Number of leaves per plant
IBA	234	1.58	4.45	2.54
$5\,cm^3$ of bacterial culture	272	1.68	5.52	3.98
$10\,cm^3$ of bacterial culture	518	1.72	8.24	4.78
None	107	0.64	1.10	1.48

(a) Explain how the scientists would have taken the stem cuttings. (3 marks)

ⓔ This question tests your knowledge of practical work from the section on cloning (pp. 45–47).

(b) Explain how the scientists could determine the number of bacteria in the culture that they had made. (4 marks)

ⓔ This question tests your knowledge of practical microbiology.

(c) Suggest what precautions were taken during the 30 days of the experiment to ensure that valid results were obtained. (5 marks)

ⓔ When you are asked a question about precautions or limitations reread the question carefully, putting yourself in the position of the researchers. Planning, implementing and evaluating practical work prepares you for this.

(d) Explain why cuttings were included that were not treated with IBA or the bacterial culture. (2 marks)

ⓔ Think controls!

e) Explain the likely roles of *A. chroococcum* in the growth of the cuttings and support your answer with information from the table.

(6 marks)

ⓔ You should recognise the name *Azotobacter* from your knowledge of nitrogen cycling (p. 58).

Student answers

(a) Cut across a stem just below a lateral bud with a sharp knife or pair of scissors. Place the piece of stem on a hard surface and make a diagonal cut so there is a large surface area for absorption of water. Remove any leaves at the base of the cutting, but make sure the cutting has at least two leaves. Reducing the number of leaves decreases the stress from transpiration.

ⓔ **3/3 marks awarded** This answer has good practical detail.

(b) Take a known volume of the bacterial culture and spread on an agar plate. Incubate at a suitable temperature for 24 hours and then count the colonies. Each colony is formed from one bacterium. If there are too many colonies to count, then dilute a known volume of the culture by a factor of 10 and repeat. Keep repeating the dilutions until there are 30–40 colonies per plate. Take replicates and calculate the mean and SD.

ⓔ **4/4 marks awarded** This is the **viable count** method, which is better than direct counting or using a colorimeter because it records all living cells, not living *and* dead cells.

(c) The cuttings should be kept in a protected environment so that they are not exposed to abiotic or biotic stress, such as overheating, pests and diseases. For example, they could be kept in a glasshouse or in a polytunnel that is ventilated and/or heated to keep them at a constant temperature. If necessary the cuttings can be sprayed with insecticide to kill insect pests. Grazers, such as slugs, snails and rabbits, must be excluded. The cuttings should all receive light for the same length of time each day and of the same intensity. If they are protected from light that is too bright they should all be shaded to the same extent. When first planted, cuttings do not have roots, so they do not take up much water. The surrounding air should be kept humid so the cuttings do not lose more by transpiration than they can absorb.

ⓔ **5/5 marks awarded** This answer covers many of the biotic and abiotic factors that could influence the growth of cuttings. This ensures that the only difference between the groups of cuttings is the treatment to which they are exposed. The answer could also explain that fungal pests grow well in humid conditions and can be controlled by using fungicides. It is unlikely that the abiotic conditions will be kept constant, so the cuttings must all experience the same fluctuations in these conditions.

> (d) These are the control cuttings. The cuttings treated with IBA and the
> bacterial culture will be compared against them to measure the effects
> of these treatments. The cuttings might respond even if they are given no
> treatment. The differences between the cuttings that have been treated and
> the controls can be tested to see if they are statistically significant.

ⓔ **2/2 marks awarded** Remember that 'control' without explanation and 'fair
testing' do not gain any marks at this level.

> (e) *A. chroococcum* may be a source of plant hormones, such as IAA, that
> stimulate the elongation growth of adventitious roots. Or it might be a
> source of gibberellins that stimulate growth. *Azotobacter* is a nitrogen-
> fixing bacterium, so it may be providing fixed nitrogen to the cuttings.
> The cuttings could use the fixed nitrogen to make amino acids for protein
> synthesis. Plants use proteins for growth, for example in producing
> enzymes and transport proteins for membranes.

ⓔ **6/6 marks awarded** This is a suggest question, so any answer that seems
plausible should be accepted. This answer draws on the information given to
make the suggestion about IAA from Module 5 and on knowledge from Module 6
to suggest the supply of fixed nitrogen.

Question 2

In genetic engineering, gene constructs often contain a marker gene, such as the
gene for enhanced green fluorescent protein (GFP).

(a) (i) Explain why marker genes, such as enhanced GFP, are included in gene
constructs. (3 marks)

ⓔ The term *marker gene* is a big clue.

   (ii) Suggest the advantage of using a gene that codes for an enzyme as a
marker gene. (1 mark)

ⓔ Here you need to apply your knowledge of enzymes to a new context.

AquAdvantage® salmon are genetically modified female Atlantic salmon,
Salmo salar. These fish have a single copy of *opAFP-GHc2*, which is a gene
construct prepared from a structural gene from chinook salmon, *Oncorhynchus
tshawytscha*, that codes for a growth hormone and a promoter from ocean pout,
Zoarces americanus.

(b) (i) Outline *one* method in which the gene that codes for growth hormone
may have been obtained from *O. tshawytscha*. (2 marks)

ⓔ You may know several methods (p. 41), but give only one, with some detail.

(ii) Explain why it is necessary to include a promoter in the construct. (3 marks)

ⓔ This tests your knowledge of transcription.

(iii) Suggest how the components of the gene construct were assembled. (3 marks)

ⓔ This takes us back to genetic engineering and also tests your knowledge of DNA structure.

(iv) Outline how the gene construct is inserted into cells of *S. salar*. (3 marks)

ⓔ As with (b)(i), you should only give the main points; this question does not restrict you to one method, so several could be described.

(c) The GM salmon are all triploid (3*n*). Suggest the advantages of making the salmon triploid. (3 marks)

ⓔ Triploid means that each cell has three sets of chromosomes rather than the usual two. How does this help? Think about the arguments made against developing GMOs.

These GM salmon are grown in experimental facilities in Panama, but as of early 2016 they had not received approval for human consumption by the Food and Drug Administration in the USA.

(d) Suggest why it has proved difficult for GM salmon to receive approval for human consumption. (4 marks)

ⓔ Plan answers to questions like these carefully. Annotate the question with some ideas first and then organise them to present an argument.

Student answers

(a) (i) The marker gene is transcribed along with the structural gene so that GFP is produced inside the cells that are transformed. When exposed to ultraviolet light the GFP fluoresces and this confirms that the cells have received the foreign gene.

ⓔ **3/3 marks awarded** This is a good, concise answer.

(a) (ii) If provided with a substrate the enzyme can produce larger quantities of a fluorescent substance than transcription and translation of a gene for GFP.

ⓔ **1/1 mark awarded** This is good application of knowledge about enzymes from Module 2. The use of an enzyme means that the intensity of the colour is not dependent on the level of expression of the marker gene.

(b) (i) An example: identify the mRNA from cells that are known to express the gene. Isolate the mature mRNA from this gene and use reverse transcriptase to make single-stranded DNA using the mRNA as the template. Then use DNA polymerase to make cDNA.

ⓔ **2/2 marks awarded** Only one example is needed, so make sure the method is clear ('use reverse transcriptase') and give some detail to be sure of both marks.

(b) (ii) The promoter is 'upstream' of the structural gene where transcription factors bind to DNA and where RNA polymerase binds to DNA to begin transcription. Without the promoter the gene will not be transcribed. The promoter determines the specific cells in which the structural gene is expressed.

ⓔ **3/3 marks awarded** The terms 'upstream' and 'downstream' refer to the direction taken by RNA polymerase. They are useful when describing transcription and the control of gene expression.

(b) (iii) The sequences are cut with restriction enzymes to form 'sticky ends'. The 'sticky ends' of the DNA fragment are joined to other fragments to make the construct by hydrogen bonding between complementary bases (e.g. A–T) and then the phosphate–sugar 'backbone' is sealed by ligase, which catalyses the formation of phosphodiester bonds.

ⓔ **3/3 marks awarded** Use your knowledge of DNA structure from Module 2 when answering questions set in the context of Module 6. Here hydrogen bonding between bases and phosphodiester bonds between the phosphate groups and pentose sugars of adjacent nucleotides are used effectively.

(b) (iv) A vector, such as a virus, would be used. The modified virus can be used to insert the foreign DNA into salmon eggs. The foreign DNA is incorporated into the DNA of the eggs so that all the cells of the next generation receive the gene.

ⓔ **3/3 marks awarded** There are other ways in which the eggs might be modified, such as electroporation or direct injection, but only one method is required.

(c) Each cell has three copies of the gene, so has the potential to produce more of the growth hormone than diploid cells. Any organism that is triploid is sterile because during meiosis there are unpaired chromosomes that cannot form bivalents, so meiosis stops in prophase. This means that the GM salmon cannot breed with wild salmon if they escape.

ⓔ **3/3 marks awarded** This answer includes the three different ideas that are expected.

(d) GM salmon could escape from fish farms and compete successfully with wild salmon and other species that feed at the same trophic levels. They could introduce parasites and diseases to wild populations of fish. They could mutate to give rise to diploid salmon that could then interbreed with wild populations. Many people are concerned about eating products that have been produced by genetic engineering, thinking they could be harmful to human health.

ⓔ **4/4 marks awarded** Answers to questions like this one need to present several ideas, not just one, however well developed and explained it might be.

Question 3

The locus of the gene that controls the ABO blood group system in humans is on chromosome 9. The gene codes for a glycosyl transferase enzyme that is responsible for adding a sugar molecule to plasma membrane proteins known as H antigens to form the antigens of the ABO system on the surface of red blood cells. The alleles I^A and I^B code for two variants of the enzyme that use different substrates. The enzyme coded by I^A adds N-acetylgalactosamine and the enzyme coded by I^B adds galactose.

(a) Where in the body is the protein coded by the ABO locus expressed? Explain your answer. (3 marks)

The base sequences of the coding regions of the DNA of the three alleles of the ABO blood group gene were aligned. Table 2 shows the base sequences of a part of exon 6 and a part of exon 7 of the three alleles: I^A, I^B and I^o. The numbers are the codons in the exons that code for the gene product. The dot indicates that there is no nucleotide corresponding to the third nucleotide in codon 87 of I^A and I^B.

Table 2

| Allele | Part of exon 6 | | | | | | | — | Part of exon 7 | | | | | | |
	82	83	84	85	86	87	88	—	264	265	266	267	268	269	270
I^A	AAG	GTC	GTC	CTC	GTG	GTG	ACC	—	TAC	TAC	CTG	GGG	GGG	TTC	TTC
I^B	AAG	GTC	GTC	CTC	GTG	GTG	ACC	—	TAC	TAC	ATG	GGG	GCG	TTC	TTC
I^o	AAG	GTC	GTC	CTC	GTG	GT•	ACC	—	TAC	TAC	CTG	GGG	GGG	TTC	TTC

(b) State what would be found between codon 88 and codon 264 in the ABO gene on chromosome 9. (2 marks)

ⓔ Look carefully at the table if you cannot immediately think of two answers. 88 and 264 is one clue in this question, but so are the headings of the main body of the table.

(c) Using the I^A allele as the reference sequence, show how the other two alleles differ from it and explain the consequences for the protein produced. (6 marks)

ⓔ This question would come with information about codons similar to Table 1 on p. 7.

(d) Changes in exon 7 are responsible for the different activities of the enzyme coded by I^A and I^B. Suggest how this is possible. (3 marks)

ⓔ You may not immediately understand what is meant by '…the different activities…'. Go back to the information at the head of the question to see what this means.

(e) Explain how the allele sequences for the ABO gene could be used to predict the phenotype of a person from a sample of their DNA collected as part of a missing person's investigation. (5 marks)

ⓔ This is testing your understanding of forward sequencing (from DNA to phenotype) and your ability to use information given in a question.

Student answers

(a) The gene codes for antigens on the surface of red blood cells. These cells develop and differentiate from stem cells in red bone marrow.

ⓔ **3/3 marks awarded** This tests your knowledge of Module 2 — the production of erythrocytes derived from stem cells in bone marrow.

(b) There would be more codons in the coding regions within exon 6 and/or exon 7 between 89 and 263. There would also be a non-coding region separating the two exons. This would be the sixth intron in the gene.

ⓔ **2/2 marks awarded** Remember that exons are the coding regions and introns are the non-coding regions. It is easy to confuse them and think that '**ex**ons' are cut out and '**ex**pelled'. Think of the 'e' as standing for '**essential**'.

(c) I^B differs from I^A by two SNPs. There are substitutions in codons 266 and 268 in exon 7. These are:

- 266: C → A changing the codon from CTG to ATG
- 268: G → C changing the codon from GGG to GCG

These are missense mutations as the amino acid coded by each codon is different:

- 266: leucine is replaced by methionine
- 268: glycine is replaced by alanine

I^O differs from I^A by a deletion in codon 87. This is a frameshift mutation so that all of the codons downstream of codon 86 in exon 6 are different, thus forming a protein that will not function as a glycosyl transferase. Codon 87 will be GTA that also codes for valine, but codon 88 will be CC– that codes for proline, not threonine.

ⓔ **6/6 marks awarded** Persevere at high-mark questions like this. So long as you know how to use the information in Table 1 on p. 7, this is not difficult. During your course practise interpreting information like that provided in this question.

(d) I^A and I^B alleles code for two variants of the enzyme — the two enzymes add different substrates to the H antigen so have different active sites. The changes to the amino acid sequences must affect the active site. The enzymes have active sites with different shapes and so accept different substrates.

ⓔ **3/3 marks awarded** The enzymes are known as A transferase and B transferase. The I^o allele does not code for a functional enzyme.

(e) Carry out PCR on the sample of DNA using primers for the different alleles of the ABO gene. Determine the mutations that are present. If the deletion is present at position 87 then the person has I^o. If the SNPs at 266 and 268 are detected then the person has I^B.

If there is no deletion at position 87 and the mutations at 266 and 268 are not detected then the person has I^A. The blood group can be worked out from the genotype, for example $I^A I^B$ is AB, $I^A I^A$ and $I^A I^o$ is A, $I^B I^B$ and $I^B I^o$ are B and $I^o I^o$ is O.

ⓔ **5/5 marks awarded** This is a good bit of detective work.

Answers to multiple-choice questions

Question	Answer
1	C
2	A
3	B
4	B
5	B

Knowledge check answers

1 There would be no nucleus and no organelles.

2 Met-enkephalin has the amino acid sequence Tyr-Gly-Gly-Phe-Met. The codons GGC, GGG and GGT all code for glycine. The genetic code is degenerate with more than one triplet coding for the same amino acid.

3 **(a)** If there is no lactose then the repressor protein binds to the operator region, blocking RNA polymerase, so the three structural genes cannot be transcribed.

(b) If lactose is present, it binds to the repressor substance, which changes the shape of its DNA binding site so that it cannot bind to the operator region. RNA polymerase can bind to the operator and transcription can occur. This only happens if glucose is *not* present in the medium. If glucose is present then it is respired

and not lactose. Another compound is required to allow RNA polymerase to bind to the operator region and this is only present when glucose is not available.

4 Transcriptional control ensures that genes are expressed only when required and energy and resources are not wasted making proteins that are not needed by cells. Post-transcriptional control allows exons to be arranged into different sequences to give different proteins from the same gene — this is known as alternative splicing. Post-translational control allows almost immediate functioning or inhibition of proteins in cytoplasm, so cells can respond quickly to extracellular and intracellular changes without having to wait for proteins to be synthesised.

5 The base sequence is read in threes. If a triplet of bases is deleted the reading frame is not affected. The loss of a triplet leads to the loss of one amino acid from the polypeptide, but downstream of the mutation the sequence of amino acids remains unchanged.

6 The deletion of the fifth base pair changes the sequence of codons after the start codon (AUG); the sequence of amino acids after Met is completely different; this is likely to code for a non-functional polypeptide.

7 The mRNA is AUGAUACGGCUUACGUUAG..., which codes for: Met–Ile–Arg–Leu–Thr–Leu–.... The sequence of amino acids is very different and is no stop codon, so the polypeptide will be l than the original.

Module 6 Genetics, evolution and ecosyst

8 The wrong amino acid would be inserted into the primary sequence of the polypeptide. This type of mutation is so catastrophic that it is usually lethal early in development. Mutations in other regions of genes for tRNA occur in mitochondria, which contain DNA that codes for their own tRNAs. These are related to some of the 150 known human disorders associated with mitochondria (p. 42).

9 Homeobox genes code for transcription factors. All these genes have homeobox sequences that code for the amino acid sequence of the homeodomain in each transcription factor that binds to DNA. DNA has the same shape in all organisms so any DNA-binding domain of a transcription factor must have the same complementary shape. Significant changes to the amino acid sequence of the homeodomain prevent the mutant transcription factors binding to DNA.

10 Mutation; the direct effects of the environment (e.g. scars); the effects of the environment interacting with the genotype (e.g. diet and height).

11 Each test cross would be carried out with homozygous recessive white rabbits, $c^a c^a$. All the other alleles are dominant over c^a, so will be expressed in the test-cross offspring. Any rabbit that is heterozygous will have offspring with two different phenotypes, whereas any homozygous rabbit will have offspring that all have the same fur colour.

12 They can be considered to be codominant at the biochemical level because both alleles code for polypeptides that function in red blood cells. As it is often difficult to determine whether someone has sickle-cell trait ($Hb^A Hb^S$) or not ($Hb^A Hb^A$) then Hb^A may be said to be dominant. However, people with sickle-cell trait can experience symptoms. Heterozygotes have a resistance to malaria that is not shared with people who are $Hb^A Hb^A$.

13 F_1: red-eyed males ($X^R Y$) and red-eyed females ($X^R X^r$); F_2: red-eyed males, white-eyed males and red-eyed females in a ratio of 1:1:2. Of the females, 50% are homozygous dominant and 50% are heterozygous.

14 The three boys have inherited the grandfather's X chromosome through their mother, who is a carrier. The fourth has inherited an X chromosome produced by crossing over, which occurred in the boy's mother. (Look out for questions on sex linkage that involve animals, such as butterflies, moths and birds, in which the females are XY and males are XX. Look out also for genes that are on the Y chromosome, such as the sex-determining gene (Sry) in humans.)

15 $\chi^2 = 5.38$. With three degrees of freedom, $p > 0.05$, so the difference between observed and expected results is not significant.

16 The two genes are partially linked. 87% of the offspring are parental types; 13% are recombinants formed as a result of crossing over in meiosis.

17 Variation is reduced.

18 (a) p (F) = 0.98; q (f) = 0.02
   (b) 3.92%, which should be rounded to 4% of the population (1 in 25).

19 A ($p^2 + 2pr$) = 0.45; B ($q^2 + 2qr$) = 0.13; AB ($2pq$) = 0.06; O (r^2) = 0.36

20 p (Hb^A) = (1210 + 390)/2000 = 0.80
   q (Hb^S) = 1 − p = 0.20
   p^2 ($Hb^A Hb^A$) = 0.64 = 640
   $2pq$ ($Hb^A Hb^S$) = 0.32 = 320
   q^2 ($Hb^S Hb^S$) = 0.04 = 40
   There are only five with sickle-cell disease (most likely due to early death of children with the condition), so the population is not in equilibrium.

21 Genetic variation between individuals of the same species is related to the different *alleles* that they have. All individuals of the same species have the same genes.

22 256

23 Four (dATP, dTTP, dCTP, dGTP)

24 DEGENERATE

25 Annealing is the formation of hydrogen bonds between complementary base pairs. Ligation is the joining of DNA fragments by formation of phosphodiester bonds.

26/27 The figures you calculate should give a straight-line graph.

28 They involve culturing microorganisms in artificial conditions.

29 3.74%

30 Detritivores 'shred' the food they eat, absorb some of it and egest the rest with a larger surface area for decomposers to act on. Decomposers break down organic material, recycling carbon as carbon dioxide and nitrogen in organic compounds (proteins) as ammonia.

31

Microorganism	Role in nitrogen cycle
Decomposers (bacteria and fungi)	See answer to Q.30
Nitrifying bacteria (*Nitrosomonas* and *Nitrobacter*)	Convert ammonia to nitrate ions
Nitrogen-fixing bacteria (*Rhizobium* and *Azotobacter*)	Convert dinitrogen (N_2) into ammonia
Denitrifying bacteria	Convert nitrate ions to dinitrogen

32 By replanting, to replace the trees that are felled. There is then no loss of biodiversity as a result of harvesting this timber.

Note: **bold** page numbers indicate defined terms.